MAYA HEATH

ENERGIES

- A BOOK OF BASICS -

MERLYN PRESS
630 S. Hutig
Independence, MO 64053

LIBRARY OF CONGRESS CATALOGING-IN-PUBLICATION DATA
LIBRARY OF CONGRESS #97-76514

HEATH, MAYA, 1948-
ENERGIES, A BOOK OF BASICS / MAYA HEATH
ISBN: 0-9651554-1-2

PUBLISHED BY: MERLYN PRESS
630 S. HUTIG
INDEPENDENCE, MO 64053
(816) 836-1182

INTERIOR ILLUSTRATIONS:
PAGES 30,34 © MAYA HEATH
PAGES 60-64 © BOB ISAAC

AUTHOR PHOTOGRAPH: TIM TOU STUDIOS

COVER ILLUSTRATION AND DESIGN: MIKE AXEN

Dedication

The seeds of this work were planted long ago by those individuals who shared their time and wisdom to pass on the principles upon which this work is based. It is with love and gratitude to them that I pass their teachings on. This book is dedicated to Jay Norman Froman who first showed me the gate to my inner self and the greater worlds beyond, Ed Peach who showed me how to look beyond the surface of words and symbols and charlatanry, and to my good friend and teacher, D. Anne Metcalf, without whose patience, humor and love none of the rest of it would have made any sense at all because she has shown me over and over that love is the greatest energy of them all.

Acknowledgments

It is impossible to thank all the people who helped make a book such as this possible. Without the help and support of my husband, Bob, and daughter, Adrienne, it would never have happened. I would like to thank most especially, Nick Nuessle for exceptional and extraordinary technical support. Dr. Tom Duke gets special thanks for keeping life bearable after all those hours at the keyboard. My greatest appreciation goes to Gavin and Yvonne Frost whose patient and careful editing and cheerful encouragement helped so very much in making this book a reality. Special thanks to David Ecton for last minute editing and type setting.

Table of Contents

CHAPTER 5
THE UNIVERSAL SPECTRUM AND COLORED STONES 83

INTRODUCTION

The New Age is fast approaching, and is bringing with it many wonders and revelations of personal expansion and fulfillment. As the millennium turns, many people are reaching outward from themselves towards a renewal and blossoming of personal and spiritual consciousness. The days of personal and spiritual isolationism are closing, marked by an opening of hearts and minds to both new beliefs and new ways of expressing old ones. Spirit and Science have begun to join hands to find new ways of validating and enhancing the newly discovered awareness. Thanks to mass communications, we are forging a new global consciousness and are fortunate as never before in human history to be able to look over the spiritual fruits of many cultures and traditions. Much of what passes for new is, in reality, very ancient, and, as these concepts are brought to light, we are finding that they have their parallels in many cultures all over the world.

Opening before us is that promised "Age of Aquarius". With it has come, first a trickle, and then a flood of new concepts and practices, not to mention books, recordings, art forms and the like reflecting this new sensitivity. It can be a wonderfully exciting to live in such a time, and it can also be terribly confusing. Like any new study or enterprise there is a new jargon to reflect these new ideas, and the ideas themselves can seem as bizarre and elitist as the words used to describe them. This can be both alienating to the casually curious and daunting to even the most diligent seeker. Without a basic body of concepts and words, it is impossible to approach any new field of interest in a meaningful way. How can you know what questions to ask or where to go to ask them?

This book is a guide through the maze. It defines some basic vocabulary, outlines some basic unifying concepts and explores a few basic techniques in order to give you a foothold to begin examining this new spirituality. When it is taken from the viewpoint of basic, unifying concepts and connections, this book presents you with a meaningful and usable survey of basic principles and ideas intended to open up the reader's own avenues of exploration. It will show you a few experiences and explain some terminology and concepts to help you sort through the vast, and initially confusing, amount of what is available and will help you to be a little discriminating about what you find.

This book is about connections. Regardless of what may appear on the surface, nothing exists in a vacuum as an isolated phenomenon. Everything connects. The Universe exists as a continuum of energies and relationships linking the esoteric with the mundane, the extraordinary with the everyday. Operating on all levels at all times, this continuum of energies provides that nothing happens as a disassociated accident or quirk of fate but as the result of natural interactive forces deriving from and proceeding with the flow of a balanced and harmonious universe. You connect with all of it, because you are an integral part of that harmony.

Beyond the words and techniques, this book is about empowering yourself as an individual by gaining access to the wonderful range of energies available to you. It is about finding ways to locate and access those resources, and how to find the connections each individual has with his or her higher self and with the Source of all things. But beyond all this, it is about the joy of living in a larger world with the strength and awareness to help both yourself and those around you achieve goals and live, now and in the years to come, the richer, fuller lives that are everyone's birthright. Your inner journey, like any form of growth, once begun has really no ending. I hope very much that you enjoy your voyage of discovery in the larger world and that your life will be enriched and renewed because of it.

CHAPTER 1
A PLACE TO BEGIN

Magic is part of our history, our culture and our awareness. The legends of the West are filled with references to beautiful and mysterious places and objects and the powerful characters who control unimaginable forces. As children we heard fairy tales and legends that filled our imaginations and dreams with these images. We pretended we were such folk; we dreamed of such powers. But as we became older, the focus of our lives shifted as we became necessarily absorbed with the demands of school, then developing a career and earning a living - putting together our lives on a physical basis. Our dreams of magic became disappointed, then laid aside in favor of more immediate concerns. There were no genii popping out of lamps to intervene miraculously in our fate - no crystal ball of visions - no rings of power - and nothing listed at the community college about Wizard 101.

Our world often demands so much of us on so many levels that it is easy to feel lacking in the basic qualities and abilities needed to cope with it all. It seems that our shortcomings are constantly pointed out to us by the stresses and demands of our jobs, our relationships, the circumstances of our lives and the trends of the current culture. It is easy to feel at a loss as to how to attract or develop the qualities and resources we need to meet these demands. Somewhere we have misplaced the knowledge and ability to draw upon or align ourselves with the forces and energies we need to strengthen and fulfill us. We are accustomed to finding answers and solutions outside ourselves by manipulating the physical world around us. The difficulty with this approach is that once the solution has become defined as something outside of and apart from yourself it is not inherently your own. This automatically moves what you most need out of your reach. The answers are "there"; you are "here". Like the tomorrow that never comes, the answers are outside where, by their very definition, they will remain apart from you and forever out of reach.

1

These empowering energies and forces are not something apart from you; they are not "out there" somewhere. They are not even something you can "go and get". Neither are they some peculiar paranormal phenomena that happen with a flash and a bang. They are the deepest part of your nature and you were born with them. They are part of you. They are very much naturally part of you and the natural world around you. They are part of everyone. Nothing unnatural can ever happen at all in this universe. All things occur within the framework of natural laws that make up the fabric and fiber of all that surrounds you. Events and abilities sometimes described as unnatural or paranormal are only those things not presently explained or understood. The forces and energies that cause them are integral elements of the fabric of the natural world that permeate every aspect and manifestation of creation. They connect all life and all levels of consciousness. Just as a river moves through many types of terrain as well as many nations, the energies you are going to examine are the single constant element that flows from the stones beneath your feet through all physical creation and unite this world with the many others surrounding it. They are the essence of your nature that you have forgotten or neglected to recognize.

These essential forces are part of the world around us, existing all the time as elements of the Universal fabric that makes up ourselves and all Creation. As part of all Creation they are also part of us as well. They exist within us as integral parts of our life, our nature and our being. In most people, however, they are sleeping. They exist in an undeveloped state because they have never been recognized or encouraged. What you are seeking, therefore, is not a means to attract what you do not have and is someplace else, but a way to awaken and develop the inherent powers within every individual that have been part of you all along. You need a way to connect with these greater forces and access their counterparts present within you.

Every drop of water contains all the elements that make up the ocean and the clouds; similarly, each individual contains, all of the potential essences and qualities that make up the Universe. In each tiny cell, there is the DNA from which an entire body can be reconstructed; on a spiritual level, each individual as the spiritual child of the Prime Source contains a part or reflection of all the elemental forces resident within it.

When you are physically ill, the natural immune system of the body is stimulated to produce antibodies that counteract the cause of the illness. Similarly, the problems you face spiritually, mentally and emotionally identify what strengths and qualities you need to awaken and develop within yourself in order to overcome them. For every shortcoming you feel within yourself, there is a corresponding power, quality, or energy already residing within you that will counteract this and move you farther forward towards achieving what you want and need in life. All you need is a way to awaken this natural immune system of the spiritual being.

It sometimes seems as if society has lost its ability to nurture its members, to heal them and empower them to meet its challenges in a healthy and positive way. You were probably never taught how to access the energies you need to be happy and fulfilled. There is a good deal too much said about the "burdens we all must bear" and too little said about ways to grow strong and able to bear them with more grace and less struggle. The important thing to keep in mind is that most people have no idea how vast and varied their true potential is. They see themselves as small and limited in power and scope. What you need is a way to expand your awareness of this potential - a method of pushing back the boundaries of your personal limitations so that you can connect with powerful elements sleeping within you that, in turn, connect with their Universal counterparts.

The study of awakening these faculties is called metaphysics - the science of what is beyond the physical. To the beginner, metaphysics can be a bewildering maze of buzz words and strange ideas having little or no relationship to each other. Knowledge of those words and ideas add a grasp of the fundamental concepts involved lead to an understanding of how these concepts relate to each other and to you. This in turn forms a workable fabric of useful information. It can be difficult to know where to begin such a study.

At the very beginning you may have only an impression, a feeling of disquiet or a tug of curiosity, an idea that there might be something out there more or better or different pulling you from the inside and drawing you forward. There are not even any questions at this point, because it takes at least some information and a little experience to formulate a question. Even vocabulary is lacking. Events and concepts undiscovered and only guessed at have yet to be given names. It is easy to doubt your own critical faculties of recognition and identification because, at the very beginning, you have only your intuition to motivate and guide you. But this feeling is definitely there, and maybe following that feeling is the most significant thing you will ever do.

These energies have always been a part of you, dwelling on the edge of your awareness. That feeling inside you is the sense of their stirring. This feeling is your need to identify with them, to recognize them in yourself and your surroundings and use their potential not only to rediscover a larger world and to make the one you live in a better place.

This feeling is called intuition - a word that literally means inner teaching. You will begin to discover how to identify and follow that inner voice; how to recognize it in its many guises; how to listen to its voice within you; how to develop its range and find how it can be your guide and teacher. It can open up a world of wonderful experiences, heightened awareness, and expanded abilities. It can eventually lead you beyond yourself and your physical surroundings to the larger world that is the source as well as the reflection of the energies you need.

It is your intuition that will guide you, not your common sense or reason, and it has always been with you. It is what led you to begin this study. Intuition is the feeling of rightness or wrongness that leads you to either do or refuse to do something regardless of what your logic tells you. Conscious logic is only a recent addition to the list of human survival skills, and, although it is often useful, it also has its limitations. Instinct/intuition is what has preserved us, both as individuals and as a species, and it is this that will show you where your connection is with the energies that permeate the many levels of creation. By following intuition, you can find what is healthy and balanced for you and what can help you live a healthier and more harmonious life. The more you listen to these feelings and attune yourself to the sources of these inner teachings, the more they will clarify and intensify to lead you into a more aware, more powerful and richer life.

As children, most of us were taught a definition of reality that only included the five physical senses. It was most important as you grew up to manage and master the physical world so that you could live safely and capably in it. This is fine as far as it goes, but it does not go nearly far enough, because it does not include (and often actively EXcludes) anything beyond those five. As a result of this physical practical focus, your more subtle senses were ignored and allowed to remain dormant. The techniques for recognizing and developing these senses were never truly developed, and therefore, you never developed a way of responding to these feelings, images, and dreams. You developed no critical faculty for evaluating these impulses as you did for the perceptions of the physical five, so these subtle senses will feel unfamiliar and uncertain at first because they are certainly untested territory. You do not know for certain just what it is that they are trying to tell you. So you must begin by looking at some ways to train and evaluate all these feelings and impressions to find out what they have been trying to tell you all along.

You live on many levels at once; on each of these levels you have senses and perceptions that are giving you information, just as your physical senses are constantly giving you information about the physical world. Your principal focus has been on the physical world and so these other levels and sources of information tend to be ignored or discounted unless they become too imperative to ignore. You have developed your other levels of awareness as peripherals to serve your physical life. They are not regarded as having their own validity or as relating to their own distinct levels of existence. They are not seen as interrelated, but rather as disassociated, separate functions: second-rate sources of data meant to aug-

5

ment the physical world. By expanding your focus to include and integrate these other levels of yourself, you become able to recognize and validate the information they are providing.

The emotions will be the first of these senses to guide you, because, after the physical, they are the closest and most familiar of these sensory levels. They have been developed by the same stimuli of pleasure and pain, achievement and failure as your physical senses; and although they tend to feel less defined, they are no less powerful and real. They can be the gateway to the other more subtle levels of perception and the key to their integration into your waking, conscious life. The mind and mental functions will come along after, because, when they are freed from the bonds of service to only the physical plane, new vistas of thought and information will become available to you. You can begin to investigate a world of possibilities beyond the physical senses and rational logic. As the emotions and the mind unfold, the world of spirit reveals itself, and with it your connections to the worlds beyond. You can begin to recognize yourself as a larger, more beautiful and more capable being than you had ever dreamed of.

The most important advice you will ever receive in this study is the simple phrase: Pay attention. Paying attention to these inner teachings is critical. Your inner voice is speaking to you all the time, giving you impressions and information on many levels from the world around you. But, because you are not used to listening, it may be dim and undefined. It will take some practice and patience to learn to trust these levels of yourself. You already know that things do not happen in a flash like in the movies, but how do they happen? Naturally - that is, within the natural order and obeying natural laws, not with a bang or even with a whisper. If you don't pay attention, you might miss them altogether. Things change with amazing subtlety as the currents shift at deeper levels to alter the patterns we see on the surface. You have never been taught how to recognize these events; and even when you begin to suspect that something out of the ordinary has taken place, you are likely to ignore it or discount it altogether. Recognition is the key, and to recognize a thing you must first notice it - you must pay attention.

Once these inner voices have been recognized, the information you receive from them needs to be acknowledged. You would never have learned anything as a child if your accomplishments had not been recognized and acknowledged. This is how you come to know, not only what is a significant event, but also what it felt like and how you made it happen so you can make it happen again and next time maybe a little better and a little easier. When you are beginning, even the smallest step is a kind of victory. One at a time these steps build into a pattern of recognition and accomplishment, strengthening perception and abilities. All you need to do is pay attention; and you will begin to see that extraordinary things are happening all the time and all around you.

This will sometimes be difficult because often you will have no one to help or advise you with it. Your experiences and perceptions are entirely your own. Even if you study or share experiences with a friend, a flash of insight, inspiration or instinct is a uniquely personal experience and will rarely be perceived simultaneously by another person, or in precisely the same way. So the difficult and sometimes uncertain part of this learning is to trust yourself and your perceptions and to believe in your inner instincts.

In the beginning, it is helpful to keep a journal - nothing terribly literary or detailed, but enough to log your impressions, your flashes or your dreams, what sort of thing you are working on and what steps you took. This is important because writing things down clarifies and defines impulses and events. It also helps anchor your results and manifest them in the physical plane by your recognition and validation. It is worth the extra time it will take in your day, and you will be surprised to find that after a while, a pattern begins to form. Bit by bit you will be mapping your inner landscape - giving form and dimension to the range of your new skills and perceptions. You will begin to assemble your own set of reference guides and definitions of feelings and experiences that can serve as an ongoing resource for knowledge, methods and tools as you acquire them. What you find ordinary today would have been amazing and hardly believable only a few weeks ago. Keeping a journal will help you to define and refine your impressions and intuition. It is impossible to say exactly how your intuition will come to you, by what series of impulses and impressions it will communicate. We may all be receiving the same impulses, but our brains do not all decode them in the same way. What comes to one person as colors, may appear as sound or temperature variation to another. What whispers as an audible voice in the inner ear of one may manifest as a dream or waking vision to someone else. You must learn to recognize and interpret your own set of internal images and symbols as you work with them.

What you are looking for is what will work for you. This alone is the critical determining factor. If you find something that interests you, by all means look into it with an open mind. If it does not work for you or if you are not comfortable with it, let it go and move on. You do not need anything more than your feelings on this. You will know what is right for you by the way it makes your heart smile or when things suddenly become clearer and brighter. Your feelings have led you this far; let them continue to guide you.

One pattern that will begin to emerge is the connectedness you have to many levels. By extending your awareness to encompass multiple possibilities, you are opening your inner eyes to wonderful new vistas. You are opening your heart and mind to a tremendous range of knowledge and experience that has otherwise been unavailable to you because you did not recognize that it existed. What you are really beginning to do is not learn something new, but recall and move into something that has always been with you. You are rediscovering your connection with many sources of knowledge, healing and information. You are finding out how to live a richer, more vital life in which your inner world resonates in harmony with your outer one as well as with the many other greater places of knowledge and energy. You are rightfully learning how to use your natural connection to these many worlds and levels of awareness to access them as a resource, just as you use and access the resources of this physical world.

Practical Considerations

On the practical side of things, you will be exploring the "larger world" and ways to work with some of its aspects as you find them in yourself and in the world around us. This is going to take some activity on your part. There will be places in the text where you will be given a guided meditation. This is as close as anyone can come to giving you "hands on" experience with some of the topics under discussion. It is a good idea for you to have a tape recorder available and at the points where guided meditations are given, you should record the text so that you can play it back for yourself while you are relaxing in meditation. If you are uncomfortable with doing this or if you do not respond well to the sound of your own recorded voice, you might ask someone to record it for you. Whichever method you prefer, it is important to do the meditations so that you will begin to experience for yourself the energies under discussion.

You will also probably notice that some of the subjects discussed in this book are not completely covered in one single chapter. Certain ideas and concepts will be introduced and the basics covered in some detail, then expanded and related to other topics, then returned to at a later point where the concepts tie in on a different level. That is because no single aspect of energy work can be encapsulated and kept apart from the whole for separate study and examination. Energy work is a field of study with many aspects and facets, each bearing on the other, that must be viewed as a whole. The point is not to see the world as a series of disassociated external phenomena or even as an endless linkage of seemingly random causes and effects, but rather as a related continuum. There is a greater wholeness and harmony that often escapes us, trained as we are in observing the world in terms of action/reaction or problem/solution. The levels of awareness, and the many worlds with which they interconnect all reflect a basic harmony, each corresponding with the others. They work together, each affecting and being affected by the others in a dimensional and transformational way rather than a linear reactionary one. This could be described as a consciousness-centered universe, and relating to it takes a shift of personal viewpoint away from the phenomenal world you have always known. This new world is multidimensional, not linear.

THE PROCESS OF MEDITATION

The human brain is filled with an amazing amount of just plain junk. We are subjected to a constant barrage of input every waking moment - radio, TV, traffic, signs and billboards, telephones and people. Noise impacts you on every sensory and conscious level. There are also constant demands being placed on you to respond to this input. You generate a lot of this noise yourself - your emotions, desires, fears, what-ifs, calculations and manipulations. Often you have so much noise going on that you literally cannot hear yourself think. How can you possibly know or manage what is inside you, if you are constantly coping with all this noise? So all this trash develops in place of genuine awareness and creativity, and an amazing amount of your energy and brain capacity is expended on maintaining this noise level.

You have been taught from earliest childhood to be responsible for what your body does - how to be polite physically and verbally, how not to damage your environment or disturb others but probably no one has mentioned that you should be responsible for your thoughts. As long as you aren't giving vent to anything unnecessarily unpleasant, the idea seems to be that your thoughts don't actually do very much in and of themselves other than serving as tools and motivators of your actions.

Thoughts have energy, have impact, have power. They are actually how you shape your world and yourself within that world. Thoughts are the actions of the inner self, or would be if you could find them under all that inner noise. What you think of yourself, your surroundings, and your interrelationships dictates from moment to moment their quality and their content. It dictates how you relate to life, how you either cope or don't cope. But thoughts are more than attitudes; they are forms of energy. The intensity and type of this energy acts upon its surroundings and in turn affects other energies it contacts. When you are blinded, deafened and numbed by a constant wall of thought noise, you are not able to perceive what energies your thoughts are creating around you and, just as importantly, what energies you are receiving from others and the world around you. When you begin to find your own inner core of quiet, you can then evaluate where your thoughts are coming from - whether from your own mind or from your daily environment. Once the source has been identified, you can adjust yourself accordingly.

Meditation is an invaluable technique for building and directing creative images and removing the negative or unwanted ones. It is the tool you use to build and explore your inner world and the "larger" world. Meditation is much more - and much less - than just thinking about something or deciding that things ought to be different. After all, if it were that simple you would have done it a long time ago. Meditation is a way of editing out the noise so that you can see your way to clarifying your feelings and thoughts - maybe you can even find out what they truly are. That sounds like a peculiar idea, but the truth of the matter is that until you begin to sort out the noise, you can't really be sure of what is truly yours and what isn't.

Meditation done on a regular basis will go a long way to enable you to find your way through this maze of chaos and, in the process, find yourself and your own individual patterns. This is a process by which you "clear and charge" yourself. This is the relaxed and powerful state of mind that leaves the disabling noise behind and moves you into a position from which you can direct and control your life more effectively. In this way you free up your access to your own perceptions, your own awareness, and your own sources of power that have, understandably, become lost in the midst of all that chaos. When things quiet down, you can begin to find yourself. Like a small sprout in the center of a seed, you can begin the process of growth and awareness. Then you can begin the process of communication, because you can know and be responsible for what you are sending. You can also begin to sort out and perceive what you have been receiving. In short, it is the door through the self and into the larger world.

BEGINNING PASSIVE MEDITATION -

Begin by finding a comfortable position in a quiet place. For the next ten minutes just allow your body and your mind to relax. This is not the time for going over your day's obligations, planning the week's meals, or rehashing the argument you had with your boss. Just allow you mind to drift. Do not actively think about anything. This is called Passive Meditation and its purpose is to allow the mind to gradually free itself of external noise to allow its true center to be revealed. As a matter of interest here, this is one of the many meanings of the lotus symbol as seen so often in eastern religious motifs. Like the seed of the lotus, the consciousness begins in the dark mud at the bottom of the pond. Slowly the sprout rises through the murky waters that appear clearer and clearer until the bud rises into the clear light of the sun where it bursts into bloom. This is a good image to work with at first. Allow the image of the lotus blossom illuminated by the radiant warm sun to rest in your mind's eye. Try not to really think about it actively; just allow the image to be there until you feel yourself relax and feel light and free.

Don't be surprised when this seemingly simple exercise turns out to be a lot more difficult than it sounds. When many students begin the practice of meditation they come headlong into the internal Ocean of Noise. This is what happens when, deprived of its usual diet of stimuli, the mind searches wildly for input. Trains of thought collide - pieces of dangling constructions, scenes from grade school, the taste of lunch last week. Your back feels stiff, you suddenly want a snack. The brain rattles on to itself trying to maintain the noise/stimulus level it has become accustomed to. Ten minutes begin to seem like ten hours. Do not let yourself be discouraged at first. This is the brain's equivalent of the junk-food junkie's craving for a candy bar. It takes time to get over, and eventually, things will slow down - perhaps not the first time you try this, but after a while. Just hold the image of the lotus in your mind and relax.

After some practice you will begin to notice a difference. This will act very much like cleaning your glasses or sunglasses. You didn't realize how murky they were until you took them off and cleaned them. You will gradually realize that you actually can perceive things, and you will be amazed and delighted by the richness and variety of those perceptions. You will become aware of the many levels on which you exist. Possibilities and potentials become self evident that you never realized existed before.

In the very beginning of this learning process, you may want to put on some soothing music to give yourself something to follow, or set out a picture of some scene that draws you away from the everyday stress. But whatever you prefer, you should do this every day for at least ten minutes. Eventually you should do it without music or external input of any kind.

Meditation is very much about the concept of now, not what was five minutes ago or what will be five minutes from now; only what is now. It sounds simple, but it is much more important than you think. Once you feel some clarity begin to take place within, you will become increasingly aware of your subtle senses and will be more in control of your inner and outer Universe.

GUIDED MEDITATION

There is also the technique called focused or guided meditation that is often used to achieve a particular purpose. It allows you to bypass your inner ocean of noise by following a text or focus provided by someone else. This is the technique used in the self-help tapes offering a variety of methods for changing or overcoming elements about your life that you feel are undesirable. Guided meditation can be a very effective tool for personal growth and change but it should not be substituted for your own personal passive meditation practice. The purpose of personal meditation is to quiet the inner self and access your own reservoirs of power and awareness.

There is a technique that you can use in conjunction with your personal passive meditation to help you achieve the goals you set for yourself. Once you have practiced with your inner quiet, this could be the first quiet exercise in investigating the power of your own consciousness. When you have an objective or goal in mind, try holding a mental image of it in your mind while letting the rest of your consciousness slow down around it. This will encourage a picture of your objective to form. This doesn't mean to actively think about it or plan its execution. Just passively allow the image to form of what it will be like when your objective is reached, then put yourself there within the picture. Use your imagination to clarify this image or alter it to suit your goal or objective. Use your imagination to form a picture or image of what you want. Do this until you are comfortable that it exists. Each time you do this, its reality will become stronger. Do not expect immediate electrifying results. Try to just allow the image to form and exist. Release it. Be confident of its reality. Making a journal of your meditations is a good way to monitor the progress of your vision. You may be surprised at the ways in which your goal manifests itself. This will also be covered in later chapters in a variety of techniques, so you will have many opportunities to practice your focus and refine your techniques.

Notice this word "imagination." It means literally "the act of making an image." This image is not necessarily a visual picture. It can be an image of sound, or odors, or tactile feelings, or emotions. Some people do not form visual mental images. Let your mind find the way that is most comfortable for you and this will be the way that works best. Please notice that it does NOT mean making up something that isn't real. What you make is very real indeed. Earlier in this chapter we discussed the power of your thoughts and how the images you build in your mind have genuine power to affect you, your life and your environment. To the degree that you know some-

14

thing is true, it will operate for you because you will be directing the power of spirit into it and causing it to manifest itself in this Universe. All right, it isn't as easy as it sounds; and developing this power, like all other abilities, takes some time and effort on your part. You must learn on a daily basis to redirect your thoughts and intentions toward being, doing, and having what you desire. You will never become a happy, productive, healthy person by thinking negative, self-limiting thoughts. Negative thoughts will only serve to create a negative world for you. Each day you should spend some time relaxing in a quiet place building the image of the person you wish to be, the relationships you wish to have, and the environment you wish to have surrounding you. This does not have to take a great deal of time, five or ten minutes in the morning and evening will do; but you should do this everyday. Thoughts have force and power in and of themselves as energy. They are the actions of your inner self. What you are creating with these thoughts has a power and reality, so choose them with care. This is the time you should remember that old admonition "Be careful what you wish for - you might get it." Perhaps it should be stated, "Be careful what worlds you build with your thoughts - you might have to live in them."

CENTERING

When a child is born, it is not aware that there is any difference between itself and others. It exists in a completely ego-centered universe that seems to have no barriers or distinctions. As it grows up, however, it begins to realize that there are separations between itself and others and between itself and the world. This is a healthy process because it recognizes the fact that though the child is part of Creation, it is also a unique individual with an identity distinctly its own. It also makes it more and more possible for the child to manipulate its world and move in it because it acquires "depth perceptions" and can now begin to explore the concept of cause and effect.

When you begin to work in the "larger world" you are very much like that child. You are not initially aware that there is anything other than yourself. As you begin to work with meditation, you will notice that you are becoming more sensitive to the world around you. These feelings will be heightened on several levels as you become more aware of how things feel inside and out. You will need to learn how to establish and maintain your personal space and how to respect the space of others. At this point, just like the child, you will need to develop a clearer sense of who and where you are with respect to the larger world. When you begin to work on clarifying your perceptions, it is also important for you to be able to discern between what originates within you and what originates outside of yourself generated by something or someone else. Knowing who and where you are allows you to be selective in sorting out what you want to be aware of and what you would rather not be bothered with now. It also gives you the ability to have respect for those other energies just as the child learns to have respect for the world around him or her. This process is called **centering.**

The center of an individual is not a point of physical space within your body. It is the spiritual locus that is the essential core identity point of every individual. This is the center of self, beyond what you do, beyond the physical body, beyond the nested energy shells of mind, emotion, and personality. The center is the point where these aspects of your self meet and connect with the Prime Force of Creation. This Prime Force is what is often referred to as the Universal Mind. Because of the connection to it, the center is the seat of both action and intuition that acts as the pivot point of all energy work. It anchors, generates, and directs all actions of power and is the source point to which they return once an act has been completed. It is analogous to the center of balance in a gymnast or dancer who moves the body out from, around, and back to the balance point in the process of a motion or exercise.

The center is also the center of volition or will. This means that it is at the center that you determine what can and can not be done to you. Nothing can be done to you without your consent. You cannot be acted upon without, first, agreeing to it and, then, accepting and retaining the energy of the act. The center cannot be affected by anyone, any thing, any act, or any extension of energy whether your own or someone else's. You can not be hurt, limited, or affected in any way at the center unless you allow yourself to be. But you can allow yourself to be pulled away from that center, that is, to operate from some other, less powerful point or position. You do this by accepting a belief or going into agreement with another point of view that you accept as being more powerful or important than your own. By accepting this belief, you allow yourself to be moved from your center. If this should happen, you can, at any time, release the considerations of others and return to your center.

This sounds simple but it can be a complex issue. There are many reasons that you allow yourself to moved off center. It may be easy to believe that someone else is stronger, wiser, and more able than you are yourself. Then it is easy to rely on that person's strength, energy, and judgement and to operate from that basis rather than from your own center that you may feel is less strong, energetic, or wise. At first it may seem like the best idea to act from someone else's determination rather than your own; but you will find that, sooner or later, you must find your own center and through it your own source of power and knowledge that will be both genuine and reliable. Then you will be truly strong.

You also accept blame, anger, and pain from others because, in some way, you believe you deserve these negatively limiting states. This is particularly easy when dealing with those close to you such as family members, friends, and lovers. You believe that because you love them, they love and respect you and have your best interests in mind, therefore, they must be correct when they assign these states to you. Accepting this can even be, however dysfunctional, defined as a part of the relationship. By returning to your center, you are in a position to evaluate and judge a situation or condition and can recognize which, if any, of these conditions are justified and which are not. You can, then, accept responsibility for your actions and do the best you can to assure the results are

17

positive and harmonious. Taking responsibility does not mean you must take on any guilt, shame or blame when things do not turn out well. It means being aware of your position in a situation so that you can change it for the better and, hopefully, not go wrong the next time. Without guilt, blame, or anger, you can see what is truly your own responsibility and what is not. It allows you to be more powerful in addressing the true basic issues and more able to bring your life and energies into balance and harmony.

Maintaining your center is very important when dealing with healing issues. It is appropriate to be sensitive and sympathetic to those who are ill and injured, but it is important to retain your sense of center when doing healing work so that the negative energies you are working with are not retained through misplaced sympathy or identification with the injured party. It is easy to become overly sympathetic to the person you are working with and go into agreement with their problem and its cause. This is allowing yourself to be moved off center. It will not do either one of you any good. By maintaining your center, you are able to perceive and examine a wide range of energy forms without being drawn into them. This means that the headache you are healing does not become your own.

This also means that the negativity generated by your relative, co-worker, or roommate, for example, does not get mistaken for and adopted as your own. You can recognize it for being what it is, and, once the nature and source are identified, you will be able to choose your best course of response. Be sure you do not retain any of these negative states; release them by concentrating on your center, recognizing them for what they truly are, and releasing the energy. Once your own center is recognized and established, you will also have a much greater range of flexibility in terms of outlook and response to any given situation. You will be able to interact without worry or concern of personal damage. The knowledge of what belongs where and to whom is an invaluable tool in energy work as in life.

By knowing your center and working from it, you work

from the seat of your intuition and are connected to the source of all light and power.

Centering Meditation

This will be a refinement on the simple passive meditation technique you have already been practicing. Begin by finding a comfortable position in a quiet place and for the next few minutes allow yourself to relax and release until you feel yourself quiet and clear. Now begin to breathe deeply and slowly. Pull the air fully from your diaphragm. Follow each breath as it enters your nose, flows down your airway, and fills your lungs bringing with it light, strength, and health. Feel the oxygen fire and vitalize the blood. Feel this warmth tingle and flow through your entire body. Release each breath slowly and fully through the mouth. As the releasing breath flows outward through your airway and mouth, feel how it carries impurities with it away from you. Feel it carrying away the tension and negativity built up through the day. The warming indrawn breath fills you, flows down the legs and into the feet and toes, warming them. It flows down the arms and into the fingers, making the tendons and muscles feel relaxed and fluid. Your torso begins to warm, and as it does you relax and release even more. As your torso continues to feel warm and healthy, you begin to feel a special warmth growing in your heart. It warms and grows until you can sense it and envision it as a point of light. It is a star in the center of your heart; it is clear pure light like a diamond. It grows brighter and stronger with every long slow breath you take until you can feel its strength and warmth spreading throughout your body. It is the key to your center. This is your awareness point of your inner door, your connection to the larger world, and your inner focus of health, strength, and clarity. Continue to breathe with this sensation. Allow the feeling of warmth and well-being to fill you. Do this until you feel comfortable and at peace. At this point you may move into a different guided meditation or return gently to a fully waking state. No matter which you choose, when you come to full consciousness from the meditative state, you will retain some of the strength and warmth you felt and the awareness of the star centered in your heart. Incorporate this feeling and awareness with your waking health, clarity, and awareness. Continue to do this on a daily basis until it becomes a constant element of your consciousness.

FINDING THE CENTER

As you continue to work with this meditation, you will begin to experience and express your center through the other levels of awareness besides the physical. Some people find themselves to be emotion-based, while others are mental- or spirit-based. These are just different ways of focussing the intuition and energy of the center just as the different energy centers of the body are. But no matter what the base or focus, the center remains uncompromised while you shift your perspective across the levels of awareness to achieve these different points of view. Each level of awareness has its own center or focal point of light, and each will feel slightly different. This may seem somewhat abstract at this point, but meditation and awareness will give you working experience. Do not worry which awareness center is which because they are all expressions of your one true center and are all connected to it and with each other by it. Knowledge and awareness of the center and its connection and identity with the Prime are the foundation of all true workings of personal power and energy.

You will soon discover where you naturally tend to center yourself. As this awareness grows, you can work with shifting your focus and examine how you are affected and on what levels by this shift. The greater is your awareness of the strength of your center and the extent of its connections, the more freely and easily you will be able to retain your balance with respect to it and be able to experience and express your energies through these different points of focus.

Extending From the Center & Shielding

There are many levels of energies of all sorts going on around you all the time. Some of them are pleasant, some are neutral, and some are unpleasant. Some are mild and others intense. Regardless of their nature, you do not need to deal with them all at once or all the time or, perhaps, at all unless you choose to. Dealing with them all at once would be like trying to live in the center of a busy intersection. So now you will need to develop ways of holding these energies at a distance so that you can ignore them or choose to relate to them in whatever way and time you yourself select. This is a kind of intensified energy shell called **shielding**. It is similar in purpose to putting shielding around electrical cables to keep their signal from "leaking" or "bleeding over" into other cables. This not only enables you to be selective about what you are listening to or dealing with, but it also keeps your work from interfering with or unintentionally affecting other people. Shielding, like centering, is developed by meditation and imaging.

Using your center as your anchor you may extend your viewpoint, perspective, awareness, and energies in a multitude of ways from it, drawing power from it and its connection to the inexhaustible universal energy source. It will always remain untouched and unharmed - serene and immovable - as you work from its base. Shielding is an extension from the center to surround yourself with a protective field of energy.

Shielding Meditation

This will be an extension of your centering meditation. Just as before, find a relaxing and comfortable position and begin the breathing exercise. Draw each breath slowly and deeply through your nose and release it through your mouth. Now begin to breathe deeply and slowly. Pull the air deeply with your diaphragm. Allow your breathing to draw in light, energy, and vitality and carry away the tension, negativity and impurities. Follow each breath as it enters your nose, flows down your airway, and fills your lungs bringing with it light, strength, and health. Feel the oxygen fire and vitalize the blood. Feel this warmth tingle and flow through your entire body. Release each breath slowly and fully through the mouth. The warming indrawn breath fills you, flows down the legs and into the feet and toes, warming them. It flows down the arms and into the fingers making the tendons and muscles feel relaxed and fluid. Your torso begins to warm, and as it does you relax and release even more. As your torso continues to feel warm and healthy, you begin to feel a special warmth growing in your heart; it warms and grows until you can sense it and envision it as a point of light. It is a star in the center of your heart, it is clear pure light like a diamond. It grows brighter and stronger with every long slow breath you take until you can feel its strength and warmth spreading throughout your body. Enjoy this feeling of peace and warmth for a moment. Now feel the light from your heart begin to expand. It moves outside, filling your body with its radiance and warmth. Now it glows in your skin, covering you with a silvery sheen. It moves farther out from you until you are surrounded on all sides by its glow as though you were sitting inside a shiny silver ball, just like a glowing soap bubble. It is light and effortless. It will easily filter out, diffuse, and reflect away from you any energy that you feel to be unpleasant or harmful. You are surrounded by this sphere of safety. You may look through it, if you wish, to examine anything you desire. You may reach out and draw into it anything you wish, but only when you intend to do so. Since you are inside it, you decide what will reach you from the outside. You also decide what the shiny mirrored surface will reflect away from you. Continue to breathe with this sensation. Allow the feeling of warmth, well-being, and safety to fill you. Do this until you feel comfortable and at peace. At this point you may move into a different guided meditation or return gently to a fully waking state. No matter which you choose, when you come to full consciousness from the meditative state, you will retain this field around you and can renew your awareness of it any time you wish. Incorporate this feeling and awareness into your waking health, clarity, and awareness. Continue to do this on a daily basis until it becomes a constant element of your consciousness.

OTHER SHIELDING IMAGES

Each individual has a different set of needs and tolerances that change on a day-to-day basis. As you practice with your shielding you will want to experiment with different levels of peace and privacy that it can provide. You may change its permeability as often as you like, depending upon what you are doing. If you are meditating or resting you may not wish to be distracted by any but the strongest of vibrations. Or you may tune it so that you will only be aware of what truly requires your attention. You may wish to make the surface "sticky" so that impressions of things will be held in place until you have the time to deal with them - like an answering machine or notice board that will hold your messages. If you are doing healing or other psychic work, you may wish your shield to admit only those elements and impressions that are relevant to the issue you are dealing with at the time. You may adjust it to modify the impressions you receive so that you get only information, not the full force of the illness, injury, or situation.

Your shield can also serve to hold things at a distance. As energy comes to you in any form - emotional, images, random vibrations - you can stop it just short of reaching you. You can hold it at a distance so that you can examine it rather than letting it overwhelm you. If you put your face directly on the television screen and turned the volume up to full blast, you would not be able to focus your eyes on the picture or understand what was being said. You must move back to a comfortable distance and turn the volume down. This is exactly what your shielding should do for you. External energies and information can be held at a comfortable distance and intensity level to be dealt with effectively. Then you can also determine what it is you are dealing with. Are these energies ones that you have created, or are they someone else's? Are they random vibrations, or something you should deal with yourself?

Do they belong to you, or should they be reflected away from you? It is important to remember that, just because you can perceive something, that doesn't mean you have to deal with it personally. You can stand inside your shiny bubble and make these decisions without being overwhelmed or personally affected by whatever comes your way. It is your space, and you can decide what will affect you and to what degree.

In your imaging, you should make a specific point to know that this shield is there with you at all times. No matter what you are thinking about or where you are, awake or asleep, you are surrounded by the silver bubble. You are always protected from unwanted influences; that creates a calm and safe space for you to be in. You know where you are. You are there and the rest of the world is outside it, just as you live inside your body and the rest of the world is outside.

CHAPTER 2
THE BODY'S ENERGIES
& THE CHAKRA CENTERS

As you continue working with your meditations and centering, you will become more and more aware of the energy currents within and around all things. Energy is the common bond of all things. It is what links us with this universe and everything in it. It is also how spirit manifests on this plane. All planes of existence have their own laws that govern them and allow their continued existence. It is how their elements are ordered and what maintains that order. Nothing can exist or operate on them without obeying those rules. This is why it is true that nothing that happens can ever be "unnatural". Nothing can go against the laws of nature; it is only that many things are unexplained. We live in the physical universe, consequently all things that exist and operate within it must do so by its laws and by means of physical vehicles that obey them.

The first of these energy patterns we will look at will be your own body's energy pattern because that is the one you are most familiar with (whether you are consciously aware of it or not) and the one that you will use most often. This is your next step. Just like the child we talked about in the previous chapter, you begin by moving outward from the awareness of your own existence to the awareness of your body. Now you will move your awareness of energies outward from your center to your body's energy fields, to examine how they actually behave within and around the body and how they connect with other energies. Just as you use your physical body to move among, relate to, and manipulate physical objects, your body's physical energy field is the way you relate to the subtler fields of energy around you operating on this plane.

THE CYCLE OF ENERGIES

Your body's energy field is similar to a tree. Beneath your feet is a large system of roots connected to the larger energy

25

field of the earth. Above you, branches stretch upward connecting you to the energies of the heavens. Like a tree, you are nourished by and also contribute to both sources. At the center is your body like the trunk of the tree where these energies meet, flow, and are expressed in various ways. At all times energy is flowing through this trunk from both ends at the same time. As it passes through, the centers in the body process, use, change, and pass on this energy.

The Earth is surrounded with its own energy field that extends from its center out into space. Earth energies can be understood as the intense physical energies of this plane that nourish and sustain us and all living things in this Universe. Without these vital currents and the bodies they support, we would have no way of relating to this physical plane. At the other end of the spectrum are the cosmic, celestial, spiritual energies of our connection to the Infinite. They are pure Life Force that is part of the pure Life Force of this universe, that also connects it and you to all other universes and planes. It is the thread that connects you as you exist on many levels at once and that brings you back to your center. Without these energies, you would soon wither and die regardless of the strength of your physical currents. You are connected to them and by them. They come together within you, and your life is the expression of their interrelationship and balance within you.

Initially these two energies are of such different vibratory rates that they are not compatible with each other at all. If they entered and immediately met in the middle, they would simply collide and little good, if not outright harm, would come from their interaction. This has been described as the "Animal and the Angel" within each individual, that, according to some philosophies, are constantly at war with each other within each individual. You are not just a straight conduit pouring raw energy in and out from both directions. Instead, there is a subtle and intricate network of energy centers that work to bring these forces into harmony. This works very much like your physical digestive tract. Raw material is put through a series of physical processes to render that material into a form from which the body can extract the nutrients it needs to survive. Different elements and nutrients are extracted at various stages of the process. As the subtle energies pass through and intermingle within the body's field, they are altered by each successive energy center through which they pass until they exit and are so altered that they are fully compatible with the pool of energy into which they flow. And at each successive stage they contribute some vital element of strength and consciousness to the individual.

The seven centers doing this cycling and processing are called *Chakras.* The earth physical energies flow upward, being refined, amplified, and speeded up at each center until they exit at the crown to connect with the white light of the Infinite. Also, the forces of the spiritual descend through the lower centers that condense and slow down their vibrational rate until they join with the energies of the Earth. Energy is retained and changed in its intensity to be either used or passed on. By means of this energy flow, you are simultaneously connected at all times both to the Higher Planes and to this Physical Universe, and you are constantly being nourished and influenced by both of them. You are also connected by it to every other thing, and within it you are an integral part of the life force permeating all Creation. The chakra system is where all life energies connect and integrate. The resulting harmonic pattern is the specific energy signature pattern of the individual - the spirit's "linking point" to the physical world; and it carries the code, as unique to the individual as any fingerprint, that is the way in which that person's destiny (that is, karma, dharma, life purpose, etc.) will be expressed.

There are many interpretations available of the Chakras, ranging from the very conventional Indian systems to the more modern rainbow. The chakras are a way of describing and expressing your psychic power centers. As you work with them in your meditations, you will undoubtedly find your own images and feelings to connect with them. As always, the way you imagine them best is the way that works best for you.

CHI & KUNDALINI

The combined flow of these energies is sometimes called the Chi by the Chinese and is the force tapped by many disciplines of the martial arts. It is also referred to as Kundalini by the Hindus. It is said that it is this force that is tapped by many Indian mystics while performing extraordinary feats of yoga and meditation. Kundalini is often described and illustrated as the "Serpent Power". It is said that in the untrained, the serpent of the kundalini remains coiled up and sleeping at the base of the spine, but as these forces become more developed, the serpent awakens and begins to rise up the spine as a cobra rises, at last spreading its hood of full awareness at the top of the head. To carry the image further, the power of the fully developed kundalini force is like the cobra's lightning strike.

The image of the serpent as power, enlightenment and awareness is seen in many cultures. The cobra circlet that crowned the Pharaohs of ancient Egypt symbolized the power of the sun god, Ra, manifesting through his royal agent. It was said that this cobra had the sharp-eyed power to determine any threat or untruth and would immediately strike out to defend its wearer. In the pantheons of pre-Columbian South and Central America, the Feathered Serpent is also a prominent figure of enlightenment, peace, and power.

A more peaceful image that is also used in many cultures to describe this powerful force is the lotus. Corresponding to the lotus image in meditation, it begins as a seed at the base of the spine, and as the student becomes more powerful and adept at the use of this force, the bud rises above the head and at last bursts into a full radiant flower. This is the meaning of the lotus image repeated constantly in Eastern religious art where figures of saints and deities are pictured sitting or standing on and/or holding the fully opening thousand-petalled lotus. The meaning is to symbolize their transcendent and serene power and state of being.

THE LOWER CHAKRAS

According to some sources, energy enters through the balls of the feet and travels upward through the legs to the pubic area at the base of the spinal column. Other sources say that this energy enters directly into the base of the spine, roughly where the tail bone connects to the pelvis, and rises upward, exiting the body through the crown of the skull. The lower three centers are those that express the animal, physical drive of survival and adaptation to change and environment; the upper three are those that translate spiritual energies into worldly expression. They are unified at the heart center.

The first chakra, called the base, is the point where the primal energy of the Earth enters the body at the base of the spine. Its color is red and it is the center of the basic instinct of survival, ruling the sexual drive as well as the will to live. This center governs the basic life functions and energy level of the physical organism. Its key word is force. An injury or impairment of this center will cause a general lowering of all vital forces within the body because a blockage here will impede the flow of energy at its entry point. The Base Chakra is sometimes referred to as the Artisan Center because this center is the first that imposes order on raw energy. This sets its nature and pattern, allowing it to be used by the other centers.

The negative aspect of this center is lust and carnality without consideration or check. The energy of the Base Chakra is balanced by the spiritual quality of the Crown. The action of the crown gives spiritual dimension to the activity of the life force, imposing a higher sense of consciousness of self and purpose and an awareness of connectedness with the larger universe. This center is balanced by the white/purple energy of the Crown/Lotus chakra. This balance is best described as the Animal communicating with the Angel. Each polarity is necessary for the individual to survive, and each is the opposite of the other - pure spirit balanced with pure physical instinct.

29

As the energy rises along the spinal column, it reaches the orange center located just slightly below the navel. This Belly Center is sometimes called the Warrior Chakra because it determines the force and quality of identity. It is at this point that the raw potential force of the Base Chakra is qualified and becomes self-aware. The term Warrior is often misunderstood. It does not mean just a fighter or killer. The Warrior nature is assertive rather than aggressive. A warrior is a person who is prepared to establish and defend the individual "territory", as a mother defending her young or an individual defending hearth and home. You might think of it in terms of "self defense", literally the defense of the self, the instinctive asserting of the strength of identity, the individualized personal essence of the individual that permits self-realization and growth. The negative aspect of this center is aggressive bullying and wanton greed. The Belly Chakra is bal-

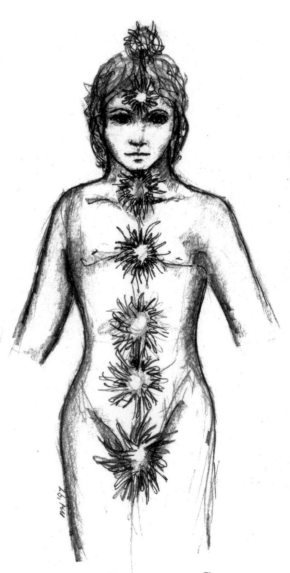

THE HUMAN BODY AND THE CHAKRAS

anced by the violet of the Brow center embodying force of intellect. It is the combination of rational intelligence with assertive force and positive defense that results in self-discipline and awareness of the self as an individual and in context and harmony with the environment.

The next higher center is located at the solar plexus. This is the yellow center called the Dancer, that rules emotions, intuition, and adaptability. It is by this center that instinct is transformed into feelings and emotions, and the individual begins to reach out beyond the self and into the world of expression. This is the seat of instinct and intuition where the self-awareness and strength of the belly merges with feelings. The solar plexus center holds the ability to adapt through change on both internal and external levels. The negative aspect of this center is emotional passion without check, reservation or compassion. This center is balanced by the Throat Chakra that is blue and embodies the energies of will and intention. The action of the Throat Chakra keeps the individual from being overruled by the emotions or swept away by whim and fancy. Adaptability is a good thing but so is the ability to control one's own relationship to the forces that surround us.

THE HEART CENTER

The Heart center is green and is called the Healer. This is where the animal drives and instincts meet and are balanced by those of the spiritual consciousness. The heart mixes and moderates these energies. You could say that the heart center is where the Animal sits down with the Angel. This is also a good metaphor for love that centers in the heart - a balanced mix of passion and adoration.

The heart is called the Healer center for two reasons. In the sense of the body's own internal regulation this is where the balance of the energy system takes place. When balance is achieved, the individual is healthy and sane on all levels of awareness - spiritual, mental, emotional, and physical. The heart also regulates and focuses the energies from the other centers for expression and use by the individual through the

31

horizontal power axis, that is, the arms and hands that are the instruments used to guide and direct energy when doing such work as healing or creating.

THE UPPER CHAKRAS

Just as primal energy is received through the Base Chakra, cosmic or higher energy is received through the crown center. Its function is the same, that is, to receive raw energy and translate it for use. And, as the lower centers connect and sustain the physical needs and functions, so the upper centers connect us to the higher vibratory ranges that sustain the spiritual life force. These energies have been variously referred to as the Higher Self, Oversoul, Godhead, Higher Planes, Cosmic or Celestial Energies. However you may wish to explain or describe it, the general idea is the same. This highest center located at the crown of the head can be called the Angel Chakra because it is the center of highest consciousness in the individual. The Crown Chakra is pure white and is what you see in religious art as a white or gold flame or flower depicted over and around the heads of saints, prophets and buddhas. The Crown Chakra is the seat of our union with the Infinite. It is balanced by the Base, for without the base to stabilize it, the physical form would not survive. The negative aspect of this center is the neglect of the connection to the physical life and lack of care for the concerns of this life. The activity of the Base Chakra reminds us that no matter how lofty and spiritual our connections and goals may be, we are still residents of a physical universe and this is the vehicle we have chosen to express our lives and purposes at this time.

As the energy descends, it centers in the violet brow chakra. This is called the Monarch center because it is here that the essence of the higher energy is translated into intellect. The intellect is then given governing authority over the activities of the actions of the individual in the universe on this plane. It is the monarch that directs the individual by a unifying vision of higher purpose combined with temporal necessity.

The negative action of this center is either tyranny over the expression of the other centers or intellectual activity without action or growth. The intellectual action of the brow is balanced by the assertive action of the belly.

The energy then passes into the blue Throat Chakra. This is the center of will and intention and is the means of expression of the personal force of the individual. It puts the intelligence and intellect of the brow into vocalization and speech. In this way the higher self communicates with the world outside the individual. It is called the center of the Seer or teacher because it is through this center that the vision of the higher world is brought into connection with the physical plane. It is the actualization of imagination. The negative aspect of this center is force of will without adaptability. Its corresponding center is the yellow solar plexus center. Just as the Dancer Center expresses and connects the self-awareness of the lower centers to the physical world, this center connects and communicates the higher energies. The solar plexus center balances the dispassionate intellectual expression of the throat with intuition and emotional understanding. Below the throat is the heart that completes the unification and balance of both energy flows.

THE HORIZONTAL AXIS

Running at ninety degrees to the vertical energy flow and crossing it at the heart center, the body's secondary energy axis runs through the arms and hands. This axis draws its power from the vertical flow and expresses it through the hands. This is the most direct and general physical way in which energy is directed by the individual into the environment.

THE FLOWER OF THE BODY'S ENERGY

To say that these energies just flow up and down and cross in the middle is very much to oversimplify the issue. The energy lattice of the body/spirit is a complex and balanced system. Imagine the human body superimposed on a multilayered flower with the heart at the center with three tiers of petals radiating in each direction. Each of the chakras harmonizes and balances with its opposite. It all works as one interlocking whole energy complex, not as a string of colored beads or lights strung one after another with no communication or relationship to each other. As we discuss healing a little later, you should remember this. If one area is weakened or damaged, its opposite harmonic center will try to compensate for the weakness until the whole system can be restored to balance, either by its own powers of recuperation or with the assistance of healing energies from outside. When you meditate to work on and strengthen one area, you must also take into consideration the effect this will have on the system as a whole. It is a good idea to do chakra meditations, with the end in mind of balance and harmony.

THE FLOWER OF THE BODY'S ENERGGIES

CHAKRA MEDITATION

In order to give you some direct experience with the chakra centers in your own body, we are going to do a meditation sequence that will lead you through these energies. Find yourself a comfortable position, preferably reclining or lying down. Now take a deep breath all the way from your diaphragm. Hold the breath a moment and then release it slowly. As you exhale, feel the cares of the day drain away from you and the tension unwind. Again breathe and exhale slowly. Feel how a pleasant stillness is growing in your mind as the outside world becomes more and more distant and unimportant. With each new breath the body and mind respond to this peace and stillness. Continue breathing deeply, and, as you do, your body becomes more and more relaxed, at peace, healthy and comfortable.

As the breathing continues and the body and mind relax, you know it is safe and right to drift away for a moment and begin your journey into spirit and the energies within you. You feel yourself sinking deeper and deeper as a pleasant comforting darkness surrounds you, cushions you like velvet. You drift pleasantly on the cushion of quiet darkness.

Slowly you begin to be aware that this darkness feels alive. You begin to feel moving currents within it. Looking around, you see shimmers of pastel light. Soft strains of music and the vague murmurs of voice whisper in your ears. The space around you is growing lighter now, and all around you is a softly luminous glow. As you pause to experience the life and texture of this place, you see before you a large mirror whose surface is dark. This is the Glass of Seeing and upon its surface can be reflected anything you wish to see. It may reflect events at any time and in any place. It may speak with many voices. It is your reference point and focus at this time. Remember this Mirror, for you may return to it at any time for these purposes.

Now the surface of the mirror begins to glow and become clear. The mist across it vanishes to reveal the figure of a human body standing erect and facing you. As you look closer, you realize that this is a reflection of your own body. See how it moves slightly as though alive and waiting at rest, for all things that are alive exist in constant quiet change from moment to moment. Now notice how this image is surrounded by a faintly shimmering glow moving and changing with the body like a shining second skin. It is like a misty halo surrounding the entire form. This is the appearance of the life current that surrounds and infuses all living things. You know that because what you are looking at is the reflection of your own body; this glow of life surrounding it is what is surrounding you. As you watch this energy closely, you will see how is seems to be flowing, passing upward through the soles of the feet, traveling upward through the legs, coming together at the base of the spine. Notice the pattern of light as it enters and travels upward through the spinal column. Watch how, as it comes together, it appears to be, not a single stream, but two streams twining together in a spiral. This is the helix of life that is mirrored in all living organisms and in all the singular elements of those organisms from the largest creature to the smallest cell.

As the flow of energy enters the base of the spine you can see how it expands and changes color. Just at the point of the tail bone you see a rosy red glow like a flower within the body. It rotates slowly pulsing slightly in time to the heart's rhythm. Come a little closer so that you can see this center more clearly and feel its energy. This is the base primal center and provides the driving force of the physical survival of the body. It is the center of both procreation and of creation on many levels of awareness. Feel its energy and warmth as it glows within your physical body renewing your health and energy.

When you have taken a moment to examine and experience the nature of this center, allow your awareness to travel slightly upward. You will notice a center just slightly below the navel. This center is glowing with energetic orange light. This center is also rotating slowly. This is the center of self-awareness, of defense, and of justice. It is the first level of perception of life outside the self and the primal drives, and it is the initial gateway to the evolution of conscious life. This is the point of nurturing. Move close to it, feel its warmth and the flavor of its light. Feel its answering warmth in your own belly strengthening, assuring, and protecting you.

When you are satisfied that you have experienced its nature fully, allow your awareness to travel slightly upward to the bright yellow flower of the solar plexus. This is the window of feelings, emotions and intuitions, animal instinct, and an awareness of the connection to all the energies around you that binds you in the intricate web of life. Its light penetrates and illuminates as it strengthens your ties with life and knowledge. It is the ability to sense and adapt to change. Feel its light sparkle within you; let it blow away the cobwebs and rejuvenate you.

When you are comfortable with your experience with this center, allow your awareness to travel slightly upward to the radiant green flower at the heart. This is the center of power, of force, and of love, where the lower energies of the survival centers are balanced and mixed with the higher frequencies of the centers of will. It is at the heart that they find understanding, expression and harmony. It is also this center that gathers these forces to change and qualify them before expressing them outward through the hands. Feel its clarity, strength, and warmth.

When you are satisfied that you have felt and experienced this center, raise your awareness once again, this time to the throat area where you can see the blue center glowing. This is the center of speech, where the intellect reaches outward to communicate with the world around you. It is also the center where the reason, will and purpose are expressed, where they become active and outreaching. You can feel its presence, strength, and confidence resonate within your physical body.

When you feel that you have experienced the nature of this center, allow your awareness to be directed upward again to the brow center. Its rich purple light strengthens and assures you because this is the seat of the intelligence and reasoning faculties. This is the place where the spiritual self evaluates and processes perceptions. This is the center that stands as the guardian and guide for you in the physical world as the

37

closest connection between your instinctual self and the higher awareness. Allow its clarity and strength to resonate within you. Feel yourself becoming even more clear and calm.

When you are comfortable with your knowledge of the essence of this center, once again you can allow your perception to move upward. This time to the crown of the head where the pure white energy center connects to the energies of the higher planes. This is your spiritual center that some sources call the godhead. As you experience this center you feel lightened and strengthened, expanded throughout. Peace flows through you and a feeling of effortless power. You can feel the light and energy of this center respond to the center in your physical body, bringing a feeling a health and strength. As you experience the energy center you will see how the current of energy continues to rise like a twisted silver cord, connecting the organism with all other planes and the infinite life essence of the Universe.

Take a moment now to move back until you can see the entire form. Study the flow of energy and the lights in the body, how their patterns of luminosity and movement complement and support one another in gracious harmony. Notice how easily and cleanly the energy flows within the system and how it passes outward through the crown, returning to the universe, forever completing the cycle of energy. Now turn your attention once again to the heart center. Notice how the energy swirls within it. As it does so you become aware of a second helix running across the chest, through the shoulders and elbows and out into the hands where once again you can see a flower of light. Only this time the lights are not one single color, but pure clear radiance made up of all the colors. Bring the palms of the hands toward each other until they are only a couple of inches apart. See how the energy streams outward from the palm of the dominant hand and how it is received and absorbed by the palm of the recessive hand. Feel how the energy cycles and strengthens as it flows through the heart center. As the power circulates, you notice how the intention of the heart changes it, colors it, makes it slower or faster, broad and gentle, or piercing and intense. Take a moment to feel how this can take place in your own physical body. Feel the current resonate through you. If you sense blockages in the flow of energy, either through the flow of chakra centers or the horizontal axis, you may now take a moment and smooth them out. Strengthen them where needed, clarify where they seem muddy, free flow where they seem clogged. Replace darkness with light - feel how this resonates with your own feelings of health and well-being. Work with this until you feel comfortable and content with what you have accomplished. When you have finished, know that the image in the mirror is your image, and that as you have clarified and strengthened its energies, your physical body and spiritual energies will benefit and become more vigorous, healthy, and more able - you will become more aware of life and all things on many levels.

When you have become satisfied with your experience with these energies of the body and spirit, you may begin to return to your physical body where you have left it resting peacefully.

You withdraw from the place of the mirror knowing that at any time you wish, you can reenter this state of awareness and the mirror will be there for your experience and use. You may use it to work not only on yourself but also on others.

You are returning peacefully now and reentering your body. You feel its breathing, you feel the blood beating in its veins. You become aware of its weight and you begin to become aware of the small sounds and sensations of physical life. You are aware of a pleasant change that has taken place. Feel the renewed vitality of your body, the elasticity of the muscles, its energy, and a new peace of mind. Remember that whenever you use this technique you will return to the physical body rested, refreshed, and renewed in body and spirit. Now you are fully awake.

LOTUS MEDITATION

The full chakra meditation is a good one to do while you are beginning to familiarize yourself with the intensity and qualities of each of these centers. It is also something you may wish to return to from time to time to balance and restore your inner energies after a period of physical, emotional, or spiritual stress. It will help you restore yourself to your own inner rhythm of harmonies. But if you wish to do chakra meditation on a daily basis for a while, you may choose a simpler form of this exercise. You can even incorporate this with the centering exercise.

When you have reached a level of calm in your meditation and breathing, shift your awareness to the base of your spine. Allow the image to form of a warm glowing seed centered right at your tail bone. Allow your attention to remain on it until you can feel the warmth coming from it. Then this feeling of warmth and light begins to grow and intensify as it begins to rise upward through your back bone. First it reaches the Belly Chakra and spreads its warmth through the lower belly area. Then it rises to the solar plexus, spreading health and light as it passes there. Next, it reaches the heart and fills the entire chest cavity with a rich comforting glow. This rises through the shoulders and pours down the arms and into the hands, filling your palms and fingertips with warmth and light. Next, it rises into the throat center where you can feel its strength warming and expanding through the neck area and up into the base of your skull. It gently expands inside your head, filling you with peace and strength that now shines out from the brow center. As the golden light rises upward from the brow center, it centers on the top of your head making your scalp tingle and relax. Now the light expands and flares, growing ever wider to form a radiant flower shape that covers the entire crown of your head. Hold this image. Allow its peace and gentle warmth to fill your entire body with golden light. Feel how all the systems of your body are energized and invigorated. Feel how your heart expands filling you with joy and peace. Feel these energies pivot gently around your center. When you come to full consciousness from the meditative state, you will retain this strength and warmth, feeling balanced and alert. Incorporate this feeling and awareness with your waking health, clarity, and awareness. Continue to do this on a daily basis until it becomes a constant element of your consciousness.

YOUR CENTER AND THE CHAKRAS

There are many varied opinions about where you will find your center in relationship to the chakras. Your center is just that - the midpoint on many levels where the Universal Energies meet and connect. The earth-based fire energy rises upward through the lower chakras and encounters the descending higher vibrational forces. The center is the pivot point at which you access these energies, and the kind of energy work or meditation you are doing will have a direct bearing on where you experience the center to be. Different individuals experience this meeting at different places. In this exercise, we have used the heart center because it is easily and comfortably accessible. Healers often use the heart for their center because this is where both essences of the main force of the kundalini energy balance and where the power is accessed for use by the horizontal axis. Others experience the center in the brow as one bright point of celestial direction where their physical universe self encounters their higher self. Some Eastern practices teach that it is the belly center. Certain martial arts disciplines use this lower center as the balance and pivot point and concentrate on drawing the earth's energy and power through it. Such a wide variety of opinions and practices is understandably confusing.

The truth is that the center is not a place at all. The center and its connection to the greater worlds exists within you all the time. It is not a phenomenon that is casually summoned when you feel you need it and then put away until "next time". Nor is it at one location to the exclusion of the others. It is best described as a point of origination and what you are actually doing is shifting yourself around it. Your body's chakra centers, mind, emotions, and spirit are the means of its expression. This is like putting colored filters in front of a spotlight. The light at the center remains the same, but the way it expresses itself changes. The purpose of meditation is to increase your awareness of its presence, its nature, and its strength until it becomes the active pivot of your actions and point of view. As an exercise of concentration, you can "center yourself" in different areas of awareness until moving around and with the center and the chakras becomes a comfortable and natural feeling.

CHAPTER 3
THE LEVELS OF AWARENESS, THE FOUR CORNERS OF THE UNIVERSE, & THE DRAGON

In your work with centering, you will encounter different ways in which your center is expressed depending upon which level of awareness is more predominant in your life at the time. There is much more to the Levels of Awareness than just a way of describing generalized viewpoints or states of mind and attitude. They describe actual energy states and levels of being in which you exist all the time. There have been many different ways of explaining and describing these energy levels. The levels of awareness exist all at the same time within you as modulations of the basic energy of your center. That sounds really obscure, but it is like playing a chord on a guitar. The chord makes a single key sound - like a C chord or a G - but that one sound that is heard by the ear is really made up of six separate strings, each one playing a different separate note in harmony with the others. The levels of awareness are similar to those strings, each playing its note in harmony with the others that make up the single key chord you recognize as yourself.

Your levels of awareness can be described as "bodies" although they are not as solid or defined as that. Many people have described them in terms of colored light, like multicolored haloes nested in layers around your physical body. This is sometimes called the *"aura"* and it discussed more fully in Chapter 6 on healing. These nested energy shells have also been called "subtle bodies" that coincide with the physical one. Sometimes they are called "etheric", "auric", or "causal" bodies. There are many terms and levels of distinction about the subtle bodies, and only general agreement about which term refers to what level. It is argued by various sources that subtle bodies are either separate shadow forms of ourselves existing in closely parallel planes, or are manifestations on the physical plane of phenomena particular to this plane that are beyond the range of our physical perceptions. This is a matter of some debate that has yet to be conclusively established.

There is also a wide variety of terminology used to describe these states - sometimes in English and sometimes in various oriental languages that are difficult for most non-native speakers to pronounce. The reason for the oriental terminology is that these states of being were first definitively recognized and described as part of certain Eastern spiritual disciplines. Some people feel that these concepts are best discussed in their original languages, arguing that Western languages do not have the idea subtleties necessary to treat these subjects adequately. It is the opinion of others, however, that it is difficult, if not nearly impossible, to think about something intelligently in words that you don't understand. It is for this reason that we will explain these subtle bodies in plain descriptive terms.

Whether these subtle bodies exist as separate body forms nested around your physical one or are merely the manifestations of different energy states of the chakras, is a matter of debate. For purposes of discussion, we will simply say that you experience events on several levels at once - physically, mentally, emotionally, and spiritually. You exist on all these levels simultaneously, although at any one time you may be more centered or focused on one than on the others.

An example of this is an athlete who can eliminate all other considerations such as emotional needs or physical pain while focusing completely on achieving the goal and the training. We have all had times when we have been concentrating so intently on a particular thing that we have neglected to notice that we were cold, tired, or hungry. We have all heard of "drowning our sorrows" in work or exhaustive exercise, and by this is meant that we concentrate so intently on a field of work or study, or physically exert ourselves to such an extent, that our emotional problems cease to trouble us for a while.

44

These are some examples of shifting the focus to different levels of awareness. This does not mean that their input ceases to exist while you shift your focus away from it. It means that you have "turned up the volume" of one level so that the others are not noticeable. These levels exist all the time and you are receiving information and impressions from them constantly, but because of the nature of your activities you may be aware only of the one you are dealing from at any given time. Each level has needs that must be met, communications that must be understood and responded to, and expressions that should be shared. If you neglect or discount any of these levels you will eventually run into problems with inner imbalances. To exist on all four levels in a healthy manner, you must be aware of your needs on each level and find ways to meet those needs that are balanced and appropriate. You feed your physical body when it is hungry, exercise, care for its condition, and are responsible for its actions with respect to others. In the same way, you must also become aware of the corresponding conditions of the other three levels of your existence so that those needs are also met and the actions on those levels are also responsible.

Each level of awareness has its primary seat of expression and function in two of your body's chakras, with the exception of the spiritual body. It seats in only one, but envelops and infuses all the others. The chakra seats are in pairs - the lower of the two chakras describes how that body relates to you, the individual; the upper of the two, how you the individual relate to others and the surrounding environment - an introverted and extroverted chakra for each, so to speak. The two chakras where each body seats are, by no means, the only chakras affected by that body. Each level of awareness uses the entire energy system as a means of focus and expression and as a means of communicating and interacting with the other bodies of the individual, enabling information interaction and communication with the surrounding environment on all these other levels. Each takes its part in governing the others and, in the balanced individual, each has a recognized voice in the decision-making process of daily life.

You are already familiar with your **physical body** and the physical universe to which it relates and corresponds. This is your vehicle in the world of solid matter and it is the body you have had the most "training" to use. This is the "locus" around which the others are centered and your primary tool for interaction with this universe. Its primary activation/motivation centers are the base and belly chakras. These are the centers that dictate physical survival, assertion of territory, and definition of individual identity.

The **emotional body** is connected through the solar plexus and heart chakras. These are the centers that govern feelings, both the instinctual, "gut level" variety centered in the solar plexus, and the broader, more articulated emotions of relationships that center in the heart. The solar plexus center connects instinctive reactions to nonintellectual and nonphysical stimuli and allows you to act upon those perceptions. The heart is the chakra with which you interact in terms of feelings and expressions of those feelings.

The **mental body** is the seat of logic and deduction, centered in the throat and brow. The throat focuses the will and the expression of that will, particularly in the physically manifested form of speech. The brow is the organ of intellect, observation, and analysis. It is the means by which you evaluate the elements of your surroundings.

The **spiritual body** envelops and surrounds the others. It is the connecting cord with the spirit that pervades and envelops all of creation in this plane and forms the path or gateway to other planes because spirit is the common link with all of them. Its single seat is the Crown or lotus chakra at the top of the head.

Although they can be discussed separately and have separate seats of connections within the chakra system, you should remember that all four levels of awareness are intimately interconnected, to communicate and interact with one another constantly.

THE LAW OF CORRESPONDENCES

The levels of awareness reflect the four basic energy states called elements in the larger world. The idea that explains this is called the Law of Correspondences. This concept is the connecting thread that ties many concepts together. There is a saying that goes, "As above, so below." It means that any thing or condition that exists on one level of existence, exists on all of the others, because all of the worlds are reflections of each other. The energy will manifest itself differently on each plane because each one has its own rules about how things can come about and exist on them. This sounds very abstract and arcane. But one of the things that it means is that, by observing ourselves, we observe the Universe. From becoming aware of how energy moves within you and how your own levels of awareness connect and interact with each other to manifest in your own life, you can extrapolate how the larger world operates and how the energies of this physical plane as well as others move, interact and manifest.

MICROCOSM TO MACROCOSM - THE FOUR ELEMENTS

Inspiration comes to a person and by a process of intuition, passion, and physical exertion becomes a reality. In the same way, energy enters the physical universe and by a process of condensing through denser and denser states, becomes manifested into material existence. You can see by the Law of Correspondences how energy enters into this Universe and distills itself into matter. This sounds like a large leap, but is really a logical process. Just as you as an individual have four basic levels of awareness, these levels of awareness correspond to sets of senses that react to and receive messages from the four levels of activity, or states of being, they correspond to. These states of being in yourself correspond to the four basic states of energy in this Universe. These states of energy exist (like everything else) at gradually diminishing rates of vibrational frequency - from highest to lowest or the other way around, depending on your point of view.

Nothing can exist in this Universe without having come into being through these four levels and continuing to exist on all four of these levels at once, just as you could not exist without all four levels of awareness existing all at once. In the ancient world, these four states were referred to as The Four Elements, referring to the four basic metaphysical components of this Universe. Simply put, this means that everything in this Universe could be described in terms of one or more of these elements, singly or in combination. They are Air, Fire, Water, and Earth and they correspond respectively to your spiritual, mental, emotional, and physical levels or bodies. Sometimes Spirit is referred to as the fifth element, as the one cohesive force that is contained in the other four and that unites them. However, it is also argued, and just as convincingly, that the four elements are the ways in which Spirit manifests in this Universe, and so it would not be accurate to speak of it as an element itself.

For our purposes here, we will stick with the traditional four. Many different cultures in all ages and from all over the world have identified these same elements and their function in the fabric of the world. We find references from India, China, Ancient Mesopotamia, Egypt, Europe, and Native American cultures, just to name a few, spanning at least 5,000 years. There are even some early Christian references to the Four Gospels being the four pillars of the Kingdom of God on Earth giving similar correspondences to directions and elements. There are naturally some differences in description from culture to culture, but the similarities are striking.

In seeking to identify and describe these four elements, peoples all over the world have ascribed living characteristics to them. For thousands of years they have been described as energy beings such as angels, demons, spirit animals, or fairy-like sprites called elementals. These elementals can be described as the existing conscious essence of the element. These visionary entities are also identified with the four primary directions of the compass, winds, times of day, and stages of life. For your convenience we have included a chart of a few of these correlations that should give you some idea of how their influences are defined. It would be impossible, however, to list all of the equivalents in all the major disciplines.

The progression of the four elements describes the stages through which energy progresses from pure essence to concrete form - from pure energy to physical manifestation. This also describes how conditions manifest in your life through your levels of awareness.

This is the point where all things begin. It is the place of ideas and concepts, and could be described as inspiration before taking action. It represents the Higher Self and the Higher Planes where the essence of all things begins. Air is the Breath of Life symbolized as Dawn or Birth. Air elementals are called Sylphs and are often pictured as winged sparkling fairies. The Archangel of Air is Gabriel.*

In Fire, the inspiration of Air takes on the first generation of form, becoming ideas. This is the planning stage of a project where the spiritual impulse of air develops into thoughts and ideas. Fire is the rational mental mind. This is symbolized as Noon or Childhood, and we begin to see the potential of the Infant as self-awareness begins. Fire elementals are called Salamanders and are illustrated as living flames. The Archangel of Fire is Michael.

Water is the element of emotions. It is what connects and communicates between the spiritual higher self and pure reason with our conscious physical world. Water is the point of transformation at which the insubstantial world of inspiration, ideas, and energy begins to become concrete and enter the world of matter, although it has not yet achieved the stability of true form. This is the point of Evening or Adulthood where the potential of the child is showing results. It also implies intuition, since it is the point where the fluidity of essential energy encounters the solidity of the physical Universe. The Archangel of Water is Raphael.*

In many cultures, the West has been associated with the Land of the Dead and the dwelling place of spirits and other world entities. Water represents dreams and visions because it is the gateway into form and matter but still fluid enough to accommodate the insubstantial nature of Spirit. Its elementals are called Undines and are depicted as mermen and mermaids, but looking much more fishy than human. The vision of the mermaid is very appropriate to this element because it is the mermaid who moves between the dark ocean depths representing the Higher Self and the Subconscious and the bright sunlit world of the sea shore of the waking, conscious mind, allowing communication to take place between the two.

EARTH

PHYSICAL

Earth is the last stage in the condensation of matter, where the essences of the other elements come together into physical reality. Earth is identified with the idea of abundance because it is here that energy is manifested as all possible forms. It is also identified with Midnight, Old Age, and Harvest because it represents the culmination of all efforts and energies up to this point. All efforts of shaping energy come to fruition here as the result of the changes brought about through the other three elements. It is the result of the change working in yourself and your life - enhanced awareness, greater knowledge or personal strength, the results of the release from the inhibiting and limiting factors that have controlled your life in this issue. Its elementals are called gnomes and appear like trolls or fairy tale dwarves. The Archangel of Earth is Uriel.

Awareness	Element	Direction	Elemental	Archangel	Age	Time	Season
Spiritual	Air	East	Sylphs	Gabriel*	Birth	Dawn	Spring
Mental	Fire	South	Salamanders	Michael	Child	Noon	Summer
Emotional	Water	West	Undines	Raphael*	Adult	Sunset	Fall
Physical	Earth	North	Gnomes	Uriel	Old Age	Midnight	Winter

   * Many people reverse the placement of Gabriel and Raphael, placing Raphael in the east as the healer and therefore in charge of new beginnings and Gabriel in the west as opener of portals. We have chosen the placement in the chart above because Raphael is often portrayed as being the "Lord of Time" and patron of journeys as well as being the Healer. He is also credited as loving laughter and having a sense of humor which would also correspond him to west as being the place of emotion. We have placed Gabriel in the east as being the angel of annunciations and therefore communications, changes, and new beginnings. Either is correct and works equally well. You should work with both ideas and discover which feels more comfortable.

THE ELEMENTS & THE DRAGON

   The one pervasive element of this and all other universes is Spirit. Throughout the ages, human beings of all cultures and belief systems have felt and recognized this unique and individual vitality. Although there have been many descriptions and personifications of it, one that has been used throughout the world is the image of a great beast - a dragon who lives under the earth and embodies the life force of the planet. The Chinese call it *feng shui*, and teach a practice of orienting buildings and elements within those buildings to be in harmony with this earth-current energy. In the West, we speak of *lay lines*, called in French *ouivres* for their subtle subterranean nature. These are lines of force or energy running within the earth that embody its life force, like the nerve channels that carry the life impulses in your own body. The Dragon is to the Earth as your own energy system, chi or kundalini force is to your body. It embodies all the elements within it as well as essential conscious life, and is a wonderful way to explore the nature of the elements through a guided meditation.

DRAGON MEDITATION

   Begin as you have before by breathing deeply and slowly. Breathe not only with the lungs but also with the spirit. Draw your breath upward from the Earth itself through the soles of the feet and into the body. With each breath, feel the energy of

51

the Earth flow into your body. Feel this and breathe this through every pore, through the base of your spine. Feel the energy tingle along your skin and within your blood as it flows through the skin. Feel it invigorate your muscles and your bones. It is warm and vibrant; it is effervescent. All the parts of your body are feeling energized and enlivened by this marvelous flow of warmth and power coursing into you with each breath you take. Allow this feeling to permeate your mind, your consciousness, and your awareness. Feel the flow of it within you. Relax and become one with it, become a part of it. It is the life energy of Mother Earth. This is the living force that binds all creatures together. This is the life of the great womb of Nature who brings forth all things. See yourself as a great tree with its roots deep in the earth. Feel the roots going deeper and deeper into the soil. Feel their solidity as they hold fast onto rock and soil down to the bones and bowels of the Earth, even unto the Center of the Earth. You are one with this Earth center and you may perceive it as a glowing bright center of primal fire, alive and vital. Your roots reach into it, surround it; you draw the primal energies upwards into yourself. You breath the life fire of creation upward into your body. It is light; it is warmth. It is pure energy flowing into you. As it flows upward, you feel your great branches reaching upward and outward, with this invigorating fire flowing along every branch and twig. You reach wider and higher, outward and upward into the heavens. You feel the life force flowing upward from the roots into every tiny leaf. The power surges through you and outward into the Heavens like a mighty river of fire and light. You exist within this flow, you channel it, you are alive with its force; you are part of its power; you are one with its nature. It becomes you, you become it, you are one thing, inseparable and within the Universe. As part of it, your perceptions begin to widen and expand. As the primal fire flows through you, you become aware not only of yourself but of all that it touches. All becomes part of your integral experience; you have but to reach with your awareness to become aware of any part of it. You are one with all that surrounds you, for you exist in the same matrix of life energy. This is the Universe in which you base your existence and these are its potentials.

Your awareness continues to expand across the Universe, merging into the starry heavens and across the infinite void. You feel very comfortable with this. You recognize that this is your natural element with which you are well suited. And as you become more aware of this, you also are becoming aware of the elements that compose it and the mechanisms by which they work together and interrelate. As your awareness expands, you are first aware of the great reaches of seemingly empty space. You examine it with interest and you recognize that what has appeared to be mere emptiness is a vast reach of moving potential. Here is energy in its free state, untrammeled by manifestation or coalescence. It contains elements of all kinds moving freely and at random, only lightly influenced by the heavier matter it permeates. You become aware of its airy motion and of its lively energy. You are aware of its brightness within and of all the unmanifested potentials it contains. You move within it; you are part of it and it moves within you as a part of you. You become attuned to its consciousness - as yet unformed - but nonetheless existent.

Now you perceive that its motion is increasing from random airy motion to something approaching purposeful speed. Its acceleration increases and becomes vibrantly energetic. The particles and potentials held within it are excited by this acceleration and begin to interact with one another, colliding, sometimes joining, sometimes flying apart, always moving, coalescing, separating. Now you join its flow and motion, you become aware of its rushing sound; you exist within it and become part of its motion and the acceleration carries you along with it. Now you become aware of its consciousness changed from before, now attaining focus and awareness. This is the primal beginning of acknowledged existence and you feel it pulsing and rushing inside you. You are one with its fiery nature, and you know that this flowing motion will precipitate all development. This is constant combustion without consumption or destruction. It is pure primal fire. This is constant change, an outward explosion of energy and it carries you mightily forward with its force.

Now your awareness is shifting once again, to the chain reactions sparked off by the ever-exploding fire. You see that because of its action, the suspended potential elements are beginning to achieve a bonding; they are beginning to realize their expression. They begin to coalesce. The form is fluid. All about you is still the constant motion and flow of particles in energy flux. They flow, they change, they achieve form, then disperse and dissolve only to form new patterns that also vanish. This is the level of constant experimentation and impermanence, of complete and total malleability. The absence of solidity allows total mobility of form and acceleration. You move within it, you are part of it, you experience the animation and potential of the ever- changing state of being. You recognize this state of ever-becoming, for it is part of you.

You move forward with it, carried by its force and activity, but as you extend your awareness into it once again you notice that the suspended elements are achieving a bonding now that is even more solid. Elements that have been previously swept along with the flow have united with each other. Look at them closely. This bonding is caused by their natures. Their energy matrices have stabilized and their identity patterns are completely formed and regular now. Even in their great diversity, they are patterned and predictable. See how some may join with others, fitting together surely and regularly, while others are cast off to seek bonding with elements more compatible to their natures. Once these bonds have formed, they are solid, fitted, regular, and complete. The motion slows. The coalescence is complete. All patterns are formed. The state known as Matter has been achieved and its manifestation is the physical Universe.

Now extend your perceptions within it, and through it. This is the path of manifestation, the path of the universe. This is the constant journey to the state of what you know as physical reality. You recognize now that what has seemed rigid and immobile is but a point in the expression of energy and subject to constant change, circumscribed only by the laws governing this place and the Universe. There is great pleasure in this knowledge, for you realize that what has seemed a prison of permanence is but an illusion. It is a constantly changing and flowing panorama of manifestation and expression. All things are malleable from this viewpoint. All things are subject to influence. Consider yourself. You are fashioned of the same stuff as all the rest. You move within it as a natural creature of this environment, for you see that on all levels of awareness you and it are the same. You are life and expression of life, within and without, spirit and form. You are one with the flow of its continuity.

Now you perceive it in all its diverse complexity and all its elegant simplicity, because you see not only the manifestation but the reality behind the expression. It is that which is called life. This is the river of life flowing with the ever-changing patterns of awareness of spirit. It is spirit becoming aware of itself and expressing that awareness. Now feel this life even deeper. It is only one life, not many. You are aware of its great pulse, of its energy flow. You become aware of it. You see before you curling through the starry heavens the form of a great creature. The Great Dragon stretches before you arching and moving with sinuous grace. Its scales are rippling with a thousand lights across the heavens. Its breath stirs in the solar winds of all the suns and all the hearts of the myriad galaxies, and moves endlessly across the vast reaches. It moves through and around all the varied forms around you and permeates all of them, becomes them. All creation is alive with its graceful power and constant motion. And as it gazes upon you, you may look into its eyes and see all the expressions and possibilities of the infinite. It is color, it is fire, it is motion, all without hindrance. There are no barriers. It is the perfect balance and convergence of all the elements. The Dragon's true name is Life. It expands before you and you realize surely that you are part of it and that it is part of you and its motion is that of a great dance.

You will remember this always. You may renew this vision at any time you wish. You have experienced the nature of all conditions of life in the Universe and the one life that is the point from which all these expression are generated. You move slowly back from the creature now. You can see him still arching across the heavens through the stars. He becomes thin and misty, shining across the sky like the Milky Way. Once again you are aware of your feet touching the ground as though you were standing in an open field gazing at the stars. This is your place, your universe; you know it as surely as you know the secrets of your deepest heart.

Now it is time to return your awareness to your physical form as it lies peacefully resting below you. You can see it stretched out comfortably. You examine it and you know that in your absence, its health and vigor have been renewed and refreshed. It feels healthy and pleasant. Your spirit slips within its confines now. You feel its regular breathing. You feel its pleasant weight and density. The skin is receiving gentle stimuli through its nerves and you can feel the motion of the air against it. You become aware of the small sounds of daily life. You notice some light coming through your eyelids.

Now breathe deeply and feel yourself awaken. Breathe deeply again and feel yourself becoming alert. Breathe once again and your eyes open. You stretch, you smile. You are at peace with all around you for you know intimately the truth of its nature and your own. Take a moment to remain relaxed and enjoy this feeling. If you feel sleepy or disoriented, take a drink of water and stay quiet for a few moments.

CHAPTER 4
ENERGIES & CRYSTALS

As you learn to find your way along your inner pathways, it is natural to look along some outer pathways for a way to explore and express your developing skills and awareness. People are natural tool users. Just as it is natural for that child we have been talking about to reach for objects to use in manipulating the world around it, it is natural to look around you for interesting objects that you can use as tools to facilitate your inner journey. Crystals make versatile and beautiful tools for exploring the many avenues of energy work. They can be powerful, easy to use, and lovely to look at. It is relatively simple to allow your eye to be captured by their jewellike facets and be drawn onward from there. But more important than their singular gifts, the study of crystals, their colors, and their potentials, naturally connects with a wide variety of fields and interests. This is one more example of the Law of Correspondences in action.

From the dawn of time people have had a fascination with shiny sparkling stones. Whether they are naturally polished by the gentle action of sand and water in the bed of a mountain stream or the most elaborately faceted diamond, their light and color have intrigued the members of every race and nation. Legends and myths about their magical properties are too numerous to recount. They have been credited with a seemingly endless variety of miraculous attributes including neutralizing poison, raising the dead, and serving as the power source for entire civilizations.

It is true that crystals do seem to have great magic and power. To begin to use those powers it is necessary to know the basics - how they are formed, what they are made of, and how anyone can work with them to harness these natural energies to enhance their life and personal awareness. The first thing you should know about crystals is that, contrary to many opinions, they are not magic. They will not give you any powers or abilities that you do not already have that are not readily available. They will, however, help you to discover and develop what you are and do have. Truly, there is far more to your potential than you have ever imagined and crystals can be an invaluable tool for exploring it.

FORMATION AND PHYSICAL CHARACTERISTICS

The newcomer to crystal work is often bewildered by the tremendous variety of crystals available. There are many types of crystals; they come in all colors and many beautiful geometric shapes. Colored gemstones are cut from crystals and crystalline formations. The basic crystal that people begin with, however, is the clear quartz crystal. There are quartz crystals from Arkansas, Colorado, and New York, as well as Columbia, Brazil, and Madagascar, and many other places. There are even quartz crystals grown by people in laboratories. It also seems that everyone you talk to has a definite opinion about how their particular favorite kind, shape, or country of origin is absolutely the best. Crystals come in such a wide variety of shapes and configurations that the beginner may wonder if all these crystals really are the same type of stone. Well, yes they are. They are all crystalline silicon dioxide and are basically six-sided.

Crystals grow layer by layer so that a cross section through the center would look like the layers of an onion or the growth rings of a tree. They are formed in hollows pocket in the Earth that are filled with a growth solution rich in the free atoms and molecules that make up a crystal's structure. When this growth medium is energized by the electromagnetic field of the Earth, the crystals begin to form on the base rock surrounding the medium. They form one layer at a time until there is either no room left in the pocket or until the growth medium becomes too poor in the necessary elements for growth to continue. The size, and to some degree the specific shape, of the crystals is generally determined by the intensity of the electromagnetic field and the geologic pressures and changes in the surrounding rock. From time to time trace elements such as cobalt, carbon, or manganese can enter through microscopic pores or fissures in the rock, changing the composition of the growth medium. These will then be incorporated into the molecular structure of the crystal as it continues to grow, resulting in layers or shadows of different colors. If sufficient trace elements are present, they will cause completely colored crystals - such as amethyst, citrine, and smokey quartz - or many other interesting and unusual effects such as shadowing, layering, and ghosting.

Crystals are also grown in laboratories for use in watches, computers, and other electronics equipment. The advantage to laboratory grown crystals is that they can be absolutely guaranteed to be atomically pure and straight, completely free from trace elements that would vary the vibration and conductivity of the crystal. They are, however, much more brittle and breakable than naturally grown crystals, as they are not formed under the great pressures under the earth. An interesting point to consider is that although these crystals can be grown quickly and with great precision, no one has yet succeeded in growing one completely from scratch. They must always be started with a tiny seed from a natural crystal or, no matter how pure the matrix or intense the electrical field, they will not grow at all.

Crystals grow by stacking molecules one upon the other in regular geometric fashion. Perfect molecules stack perfectly, while molecules containing trace atoms of a different element will stack just a little differently. Roughly like magnets, one end is positively charged and the other negatively charged so that they build up in order to form one tiny layer at a time. As they build up, they form a structure that is similar in shape to the molecules that form it. This structure will have a specific number of sides in a particular configuration and at least one end, called the *termination*, shaped in a unique and distinctive way. This is the end of the crystal that grows away from the rock. Like all other natural things, no two crystals are ever exactly alike, but all crystals of any one family will be similar in the number and configuration of the sides and in the shape of the termination. In the center, all crystals have a primary center axis with secondary energy axes radiating outward from the center to the angles between the sides. It is this internal axis structure, sometimes called a *lattice*, that gives the crystal its unique vibratory rate and therefore its power.

As we have mentioned, it is the forces in the Earth - that is, the heat, pressure, electromagnetic field density, richness of the growth medium, position on the Earth's gravitational field, and geologic stability - that determine the way a crystal will look. Within the limitations of each crystal type, there can be a wide variety of external shapes, and of side and angle patterns (called *interfaces*). As these external shapes vary, so will the lattice patterns vary and, therefore, the crystal's energy potential will change. This makes a big difference in how the stone behaves and feels when you work with it. The study of crystals and their energy metaphysical properties is relatively new, and a variety of terms have sprung up outside of traditional scientific terminology to describe the variations in their structure and the resulting differences in their energy behavior. We will try to give a few general and basic definitions, although you should bear in mind that this field of study has grown so fast and so widely that terms may vary from place to place.

TERMINATIONS - SINGLE AND DOUBLE

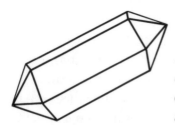 Most crystals have one natural "finished" end, having grown straight outward from the base rock, that is called a termination. Their negatively charged "intake" pole is at the base and their positive "outflow" pole is at the termination. Some crystals, however, have two terminations - one at each end; these are referred to as "double terminated". They have the seed point at the center and grow outward in both directions. The layers of a double terminated crystal form around the whole piece in one solid skin. True double terminated crystals have only one central axis that runs between the two points. This gives a very balanced and centered energy flow.

TWINNING

Twinned crystals are two different crystals that share a common base but grow at angles to one another. They may be closely similar in structure or very different from each other. Some grow directly opposite from one another so that they are difficult to tell from a double terminated crystal. Others grow at angles to each other. Twins may also grow side by side closely together, sharing a complex internal axis structure, so that neither appears to be a complete crystal - as Siamese twins share an interrelated physical structure. When crystals are twinned they can be very strong, amplifying each other, but when working with them it is important to consider where the true axis is located. If you are drawn to one of these, take time with it examining its faces and inner angles. Take a little extra time meditating with it to find its strengths and weaknesses.

GHOSTING & LAYERING

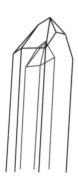

As a crystal grows in the earth, there are times when the growth medium changes its chemical composition. When trace elements are introduced into the growth medium or when the coloring element is intensified briefly, this causes a single layer or group of layers to form with a different molecular structure and color than that of the basic crystal. When these elements are exhausted from the growth medium, the crystal as it did before, and layers of clear crystal then overlay the atypical ones. This results in a crystal that, when viewed from the side, appears to have all or part of a second (or more) crystal or crystals nested inside. There may be many layers of varying intensity, or only one looking almost like a shadow. These layers are known as *ghosts.* These serve to give the crystal virtually the strength and identity of two or more separate crystals growing together and acting simultaneously as a single unit.

Beta & Tabular Formations

The term "tabular" comes from the same word root as table, that is, flat. Tabular crystals are flat on the side (not squashed from the top), looking almost as though they have been pressed in a book. They are also sometimes called "tabbies". Formed under tremendous pressure, tabular crystals are hard and dense and tend to break less easily than ordinary crystals. Their center axis is different from more regularly shaped crystals, being actually shaped like a flat surface rather than a line. Their energy is very centered and concentrated; because of this, tabular crystals are powerful for a wide variety of uses - from meditation to healing.

Beta crystals are also very flat but are much more irregularly shaped than tabbies. At first glance they may even have the appearance of flakes or broken shards. However, on closer examination they have the distinct faceted sides of their more regular kindred although the number, and arrangement of those sides can vary widely. Their energy, like that of the tabbies, is very dense and broad across their center axis. They can make strong, effective and interesting working tools.

Compressions Crystals

Another form of unexpectedly irregular crystal is called a *compression point*. As a cluster of crystals forms, sometimes crystals toward the inside of the cluster become crowded or trapped by the crystals surrounding them. This causes the shape of their sides to be dictated by the shape of the crystals they have grown against.

LASER POINTS

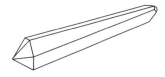

Laser points are long, tapering toward the termination and very pointed. Their sides are often extremely regular and geometrical, giving them an almost artificial or cut look, and their interior is exceptionally clear (although their outer skin may look frosty). Sometimes their sides are bowed or curved. The center axis is long and straight, giving great focus and intensity to the energy output. This is the origin of the term laser point, since it is said that they can be focused to pinpoint accuracy like a laser when used for healing or other high-intensity work.

GROWTH LINES

Although a crystal may appear to be perfectly smooth and clear, its sides are lightly striated with growth lines that run at 90 degrees to the center axis. At least one side will show these lines when you turn it in the light. This natural marking is one good way to tell a natural crystal from one that has been cut and polished.

PYRAMID MARKS

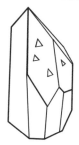

Sometimes one or more faces of the termination will have marks in the shape of triangles or pyramids. They may be deep or subtle, raised or indented. Some sources refer to these as being the identifying marks of *"teaching crystals"*. By this they mean that these crystals are thought to have some special purpose to achieve or message to impart, which has been programmed into them since their inception. Others believe that they have the unique power to bring out and teach particular life lessons to the holder, to bring out and focus knowledge and skills from previous lifetimes that may be dormant in the holder but that will have particular value and pertinence in the current lifetime.

WINDOWS

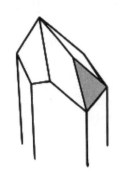

A triangular or diamond-shaped face on the termination at the base of two of the ordinary regular faces is called a *window*. Sometimes crystals with this characteristic are called *memory* or *record-keeper* crystals because some people believe that they have a greater storage capacity than regularly terminated crystals. They are highly regarded as focus crystals for trance work and meditation, because they are thought to have the ability to draw the holder's spiritual energy and focus it toward the desired result.

There is an amazing and wonderful variety in the way crystals form and grow. Like flowers, although they seem at first glance to be identical, no two are ever exactly the same. It is also impossible to classify and clearly define all the subtle variations in their formation that you will encounter. Many crystals have a variety of combined formations, making each one unique and special in both its beauty and its energy characteristics.

CRYSTAL MEDITATION

Although there are many methods and techniques for meditating with your crystal, here is a good one for directly experiencing its energy and inner nature. Choose a quiet place for this where you are not likely to be disturbed or interrupted. Make sure that you are not hungry or thirsty and that you are well rested. Find a comfortable position, preferably seated. Hold or place the crystal so that your eyes focus easily and naturally on it. Your hands and arms should be relaxed.

When you are comfortable, close your eyes and begin breathing slowly and deeply from the diaphragm. With each deep breath allow yourself to sink deeper into this relaxed state. You should be able to feel the tensions of the day gradually release from your body as you relax more and more deeply. Imagine your breath as a gentle breath of wind cleansing your body and mind as it passes through you. As you float with this gentle wind, you begin to see an open plain of clouds filling your view. Watch the clouds as they move and glow with their inner light. Allow yourself to drift with them until you are at peace and one with them. ---

As the clouds roll together, you begin to see a more intense light glowing within them. It glows brighter and brighter until the clouds roll back and reveal within their heart a brightly glowing fire: the essence of their energy. Now, the clouds begin to dissolve altogether and draw away, and you see emerging from them a glowing crystal form. ---Look at how perfectly clear it seems. Allow this perfect clarity to fill your senses. --- Feel how harmonious and balanced all its angles and planes are, and how symmetrical are its terminations - It makes the ideal container for the flame heart inside it. ---This is the flame that does not consume itself but is always perfect in its brilliance and radiance. As you see how truly brilliant it is, you feel it drawing you gently toward it and into itself. You feel it calling to that element in you that is like itself. You begin to sense and experience the kinship of your natures. Watch how it slowly turns in space, showing you every aspect and angle of itself one face at a time. See how the light of the surrounding energies reflects on its surface - and as it interacts with this light, see how these energies are part of its nature and substance. ---

Now you see little tendrils of this light emerging around it, slowly forming spiral patterns interlacing with one another. This is the manifestation of the pattern of its nature. You feel an inner understanding of these patterns, for your patterns are like them and in harmony with them. You feel the peace of this harmony and the reaffirmation of kinship and life. For this is also a center on the River of Life, just as you are. You and the crystal are part of one another. And as you sense this and understand it, you move completely within it. ---

65

Now from within the glowing crystal you can intimately experience this essence and source of its nature. First feel the center axis extending from point to point, from termination to termination. This is the spine that supports the flow of its energy, just as the spine of your body supports your body's energy flow. Now, as you feel the center axis of the crystal, you become aware of the axes extending from it to each angle of its sides. Feel how they radiate outward from it like the spokes of a wheel, forming glowing planes of energy throughout the structure. Look above and below you and see how they meet and join at the terminations. They are pulsing with life and energy, and as you feel them glowing you also feel how they are filling you with this life energy. As they glow and pulse, you begin to hear a tone, a singing, a resonance that you can hear both with your ears and from inside your heart. It sounds as though the wind has passed over the single string of a great harp and this string is the center axis, and this center axis is your center axis as well. This is the song of the stone; it is unique to that single archetypal stone. It resonates throughout your entire being and brings with it peace and great joy and harmony. As the song fills you, so the spirit of the stone fills you and you become more intensely aware of the crystal form. Look how this is the basic and most perfect form of stone. This is the singular building block of which true physical stones are made. This is what you call the cell or the seed, the unique singular basic structure of which all organisms are made. Now, you become aware of how easily and harmoniously they may be joined one with another, each singing the same true tone/song, coupling its life energy with another. This is the elemental building block that you may call a molecule, built of purely harmonious energy - complete within itself. It is a whole unit joined to form a greater whole. Like it, you too are formed from these units to make a whole; like it, you are bound together with the same life energy.---

Now you begin to move out and away from it; you once again become aware of your own distinct identity as apart from the crystal form. You are whole and filled with the knowledge of its nature. ---As you move back and away, you see the glowing clouds move toward it they flow around it ---finally hiding it altogether. ---You know it will always be there within those flowing forms when your spirit asks renewal of its energy.

Allow your body to take part of its life-giving energy and strength. Know that as you gently return to your physical body, you will feel refreshed, energized and renewed. ---You become aware of your physical body as you settle once more within it. ---It surrounds you and you are comforted by its familiarity. ---It is healed, refreshed, ready for your return. Now you become aware of its breathing, the flow of life energy within it, uniting it with all things. ---The sounds of your world return, the smells the feelings and sensations. Your eyes open slowly. Breathe deeply and release the breath slowly. ---Breathe again and feel all systems awaken. ---Breathe deeply a third time and you will be completely awake and refreshed. You are ready to use the knowledge of this experience, fully aware of what you have gained.

WORKING WITH YOUR CRYSTAL

Crystals are wonderful aids in discovering a whole range of abilities, awareness and perceptions that you had previously not been known existed within you. They can help you find the doors to many rooms to wonder and knowledge that have been resting within you waiting for a way to find a way to open into your conscious life. You should know, however, that your crystal will not endow you with any quality, power, or ability that you do not already have. There is a trick to this statement, though, because there is so much more to you than you can possibly imagine. There is a wonderful wealth of health, joy, inner strength, creativity and power in everyone. The challenge is reaching the levels of your mind and spirit where this force is located in order to tap and release its potential.

Many limitations are caused by self-imposed blocks or considerations. You have cultivated these negative thoughts, emotions, and self-images for years but were probably unaware of either their existence or the extent of their power over you. Crystals can be valuable aids in discovering, dealing with, and neutralizing the negative and self-limiting concepts and attitudes that keep you from recognizing and achieving your fullest potential. They can also help you replace these concepts with healthy, positive new images, attitudes, and goals that you can then use to enhance your own life and the lives of those around you.

We live in a physical universe; so it is only sensible that, no matter how lofty or spiritual a thing or practice may be, it must have a physical mechanism or vehicle to be able to manifest and operate on this plane. In fact, this is not only sensible; it is one of the great universal laws. Although crystals can be wonderful metaphysical tools, they must work with you in a physical way. The way crystals work with you is very much the same way in which they work in a quartz watch. When an electrical current is passed through a crystal, it emits a vibration or oscillation at a very specific frequency. With quartz crystals, this frequency is divisible by 60 cycles per second. This is why they work so well in watches: because this signal can then be broken down into a regular pulse that gives us 60 seconds per minute. Granted, that's very simplified, and it requires some complicated electronics to make it happen, but basically that's all there is to it. The human body generates an energy field around itself that can be measured in terms of current. When you put a crystal in line with this current it behaves just like it would in the watch - it oscillates at a rate divisible by 60 cycles a second. Now how does this help you?

Human brain-wave patterns can also be measured as a vibration or oscillation frequency. The level of brain wave activity at which you are the most intuitive, sensitive, and creative, which is also the level at which the brain functions during meditation, is called the alpha frequency and, interestingly enough, is also divisible by 60 cycles per second. Research has shown that it is in this state that creative activity takes place in artistic people. It is here that the body does its greatest healing and rejuvenation work, not only when recovering from injury or illness but on a daily basis to retain the body and mind's vitality. This is the level at which you dream and have visions. It is also the level that seems to generate psychic phenomena. This is also the level at which the mind is the most suggestible. This strong and creative alpha-level state is the gateway to programming your inner self and accessing your potential

We discussed earlier how thoughts have power and how they can and do shape the reality around you. To the degree that you believe something and focus your attention on or to-wards something, so it will exist and manifest itself. You do this all the time without being aware of it. The attitudes you hold about yourself and others, the point of view you hold about your surroundings and your life, constantly shape and influence your reality. Thoughts can be described as energy and force. Crystals are an ideal tool for working with energy because their output is almost pure vibrational energy. It is a relatively simple step to work with a crystal to manipulate your own energies and those surrounding you. Whether they are the energies of your emotions, your body's health pattern, or the larger fields of the events around you, you can consciously shape your life and environment. The key word here is *conscious.* The most insistent and ingrained attitudes and forces do not exist in the conscious waking mind at all but in the subconscious. You may not even be aware of them, but it is these considerations that have the strongest influence over your life.

By working with your crystal during meditation through the alpha state, you become able interact with your crystal. Its basic energy pattern is accessible to you at this level. You can communicate with it and, by doing so, you enable the stone to interact with you on the most subtle level, at which you are both the most flexible and powerful. By focusing your atten-tion on a goal or desire while you are in a relaxed and medita-tive state, you can program a crystal to reinforce this intention by charging the stone by means of your body's energy field. This activates and programs your crystal to help you to repro-gram and redirect your inner energies into positive and pow-erful patterns that can reshape your life. Then, while you go about your daily life, waking and sleeping, your crystal vibrates in the special pattern that you have set for it, guiding your subconscious mind and the energies within and around you to redirect and stabilize in the patterns you have specified. Once programmed, the crystal plays its subtle song continually. Since it is the nature of physical organisms to adapt themselves by taking the path of least resistance, when your thoughts or energy states deviate too far from the desired level or frequency, the gentle pressures produced by this difference will cause your energy levels to readjust themselves, seeking to minimize this irritation.

CHOOSING YOUR STONE

As you may have gathered from the discussion of shapes and growth patterns, there is an almost infinite number of configurations, shapes, and sizes that crystals can form as they grow in the earth. Each crystal, whether single or multiple point formation, has its own distinctive energy lattice pattern. There can be an amazing and sometimes bewildering variety to choose from when you are first looking for a stone to work with.

No two people are alike. Neither are their brain-wave patterns and neither are any two crystals. When we speak of finding "your" stone, we mean finding the crystal that is most agreeable and harmonizes best with you and the purpose you have in mind for it. Each person's alpha patterns are just slightly different; like fingerprints, there may be a lot of similarities, but each pattern is unique. The same is true for natural crystals. No two stones are ever identical and neither are their vibration rates and patterns. So, what you are looking for is the crystal that comes closest to or harmonizes best with your personal pattern and the goal you have in mind. The nature of the goal is entirely up to you. Indeed, you may not have a specific goal other than finding a crystal to wear that is harmonious with you "at your best" - a touch stone to bring out the best in you. Or you may have a specific task in mind that you have specifically worked out in detail. Or it may be something in between.

This does not mean that you must search the world over to find your one and only absolutely perfect match. As you work with crystals over time, you will probably find many that resonate harmoniously with your pattern and your goal. Your own inner pattern of energies will differ depending on your state of mind and personal development. The demands of your project or goal will definitely have a bearing on the type of stone you choose, the intensity and clarity of its energy. There is no one perfect crystal for any particular task, but you will want to look carefully for the one that feels most in harmony with yourself and your goal.

You should be prepared to take whatever time you feel you need to choose your stone with care. Some people find that it works best for them if they have a specific goal in mind when starting out to choose a stone. They hold this purpose clearly in mind as they examine the crystals available. Other people prefer not to have anything specific in mind. They look at stones and select them on the basis of how they feel about thier energy. You may be attracted to a wide variety of configurations, or may be drawn to one in particular. This may all sound very vague, but choosing a crystal is a very personal experience and differs from one individual to the next. You may find that the "one" is the first one you touch, or it may take you a while. There is no foolproof, one-and-only way to pick your stone. Some people just "know" which one; to others it has a distinctive visible twinkle that draws their eye; still others have to physically touch them all. We have heard of some people who say that they can "hear" the stone either chiming like a music box, singing, whistling, or otherwise calling to them in some way. Some say that they can feel their fingertips tingle; for others the crystals grow either very warm or icy cold to the touch. Still others say they cannot put a stone down and that no matter how hard they try to leave it alone to look at other pieces, it will always "somehow" find its way back into their hand. You may find that nothing unusual or spectacular happens at all except that you keep picking up the same crystal over and over even though you don't consciously intend to. The overall method here is that when you start looking for "your" stone, allow yourself to be open to your perceptions. Be aware of what you are feeling when you come in contact with the stones. Let them talk to you in whatever "language" you are ready to receive, and you may be surprised at what you feel.

CLEARING & CHARGING

When choosing your first crystal for the purpose of this work, we would suggest that the crystal you begin to work with be a fairly simple one with a straight center axis and fairly regular sides. At some time in the future you may wish to experiment with the more unusual forms to find out how they work for you. However, when you are first learning, we would suggest that you choose a crystal that is simple in structure and relatively clear and free from inclusions. This will give you a good experience of what the basic crystal energy feels like to work with. As you become comfortable with the work, you can to expand your experience of crystal types and the variety of energy qualities they can produce.

Once you and your crystal have found each other, you will need to clear it before you program it. A crystal retains and emits energy in two very distinct ways. A stone's basic vibration rate, its basic nature, is determined by the vibration rate and qualities of its center axis. This is unique to that crystal and does not change. This is what you were initially sensing when you chose the stone. Its "memory" (that is, the record of its experiences, programming and activity) is held in its surface in the degree and pattern of ionic bonding of the atomic structure, and it is this that you are going to clear and bring into alignment with you.

Very rarely will you be able to know how your stone was taken from the earth and whose hands it has passed through on its journey to you. In its long life it has been subjected to many random natural stimuli - intense cold, heat, pressure, electromagnetic and gravitational forces - and its behavior will have been affected to some degree by all of them. You do not want the stone's work to be muddled by old accidental patterns that you did not intend to be there. So the first step in working with your crystal is to clear it. When you clear a crystal, you are realigning the surface ionization in preparation for charging your stone with its particular purpose or message. This is similar to formatting a computer disc prior to storing information on it. Not only is the surface cleaned of old unwanted information, but it is organized in such a way that it is now compatible to the way in which the computer (in this case you) will store and use the information to be placed in it.

There are several methods for doing this, and your choice will depend upon your level of expertise, the purpose you have in mind, and what you know of the past history of the stone. We have sometimes heard of "salt clearing". This is the process of submerging your crystal in salt water (often sea salt is specified) for some period of time. This will completely remove all surface ionization from your stone that effectively reduces it to a null set. This is not generally recommended. It will remove every bit of personality or purpose from the stone by wiping it to a blank. This is the equivalent of a lobotomy or total amnesia in a human being, and in all but the most unusual cases this process is extreme and excessive. Unless your stone has been, for example, the most prized possession of an ax murderer, such a procedure is not only unnecessary but will remove many of the qualities and much of the "personality" that you found attractive to begin with. Should you come into possession of a stone that you feel cannot be cleared by more gentle methods, we would suggest that you put it away and find another stone of a less complicated and more benign nature to begin your work and use in your meditations.

If you wish to use a non-personally interactive method to clear your crystal, we suggest that you leave it in a sunny window sill in a quiet room for a week. This will allow some of the random charge to dissipate. If the stone is heavily charged, it may be necessary to leave it in the window sill or bury it in the earth for one full moon cycle. If you prefer, you may use running water. Just let the cool water from your faucet rinse the stone. This will sometimes not work very well with heavily chlorinated and fluoridated city water, but water that has not been extensively treated usually will work just fine. This reduces the surface ionization while leaving its character intact. A variant of this is to take a bath or shower while wearing your stone. We do, however, believe firmly that the active, personal method we are about to describe is generally the best.

Sometimes you may find a stone with a unique and special pattern of energy that you do not wish to change. Given the rich variety of circumstances and people that crystals pass through on their way to the local rock shop, it is not surprising that occasionally one comes along that has such a special vibration of its own that you do not wish to change it. It may also happen that a stone will stimulate or trigger a particular pleasant or interesting response in you that you were not expecting. In either case, clearing the stone would do a disservice to the special energy relationship its presence offers. You should not hesitate to wash it if it becomes dusty or to place it in a sunny window where you can enjoy its inner and outer beauty. There are many ways to interact and work with stones and sometimes just allowing yourself to participate in its special ambiance is the best.

The quickest and most effective method to clear a crystal prior to working with it, we call "impulse cleansing". This is very much like going to a pet store to find a puppy or kitten. There may be many there that are wonderfully cute and fuzzy, but when you pick up that special one, your eyes meet, your hearts touch, and something inside you says, "Mine". It is like wrapping your heart around it with one surge of energy through your hands - you "love" at it. By doing that with a crystal you both clear the stone and align it with your personal energy pattern. We use the word "love" because that comes closest to describing this feeling in words. What is actually happening is that you are focusing your energy pattern through your heart chakra, which then flows through the horizontal axis of your arms and hands. There is a long and involved explanation about the nature of focusing the heart chakra, but in terms of working daily life and reality, the process is just as simple as "loving" at your rock. That's what it feels like when your whole self, in all its aspects, unifies to one moment and act of singular purpose and intention. This is done with your hands, your body's energy field, and your focused attention. Find a quiet place to relax a moment with your stone, holding it gently between the palms of both hands. Take a deep breath and, as you exhale the breath, allow any tension you may be feeling to leave your body. As your body relaxes, your mind also relaxes becoming clearer and more focused. Now, look into your stone and envision it filled and engulfed in a pure white dazzling light that drives out everything else. Picture in your mind pouring this brilliant light into the stone with one pure focus of your will and heart into one burst of thought such as, "Mine!", "Clear!", "Clean!", or whatever feels appropriate. This will establish a clear connection between your energy field and that of your crystal. The next step is to charge or program it to your specific purpose.

Charging the stone is done in much the same way. This time, once you have relaxed and cleared your mind, consider what you would like to have your crystal help you do. At first you should keep your goal fairly simple and realistic. You might want to program a more positive self-image, inner strength, greater calm and presence of mind, or more will power to change an undesirable fear or habit. You might want it to help you develop your perceptions or abilities. You might want help in mastering a physical condition - such as overcoming the effects of an injury. Whatever it is, choose just one thing and define it clearly to yourself. With your mind's eye build an image of what you want until you are satisfied that it is exactly the way you want it. Or, if you are not the sort of person who images clearly, explain verbally what you wish to accomplish. It is often effective if you can do both. Remember to keep it clear and simple. Then, when you are sure you have it just right, you set it into the crystal with the same sort of impulse you used when you cleared it - one straight shot from your heart through your hand. You may think or even say aloud, "Yes!" or "Do it!" Again, you should experiment until you find whatever impulse trigger works for you. However you phrase it, the actual programming is done with that one strong surge.

Now you can begin the process of working with your stone to achieve your programmed goal. It will not work a magical transformation while you just stand there doing nothing. Nor can you ask for such a change to take place then ignore or resist the crystal's influence, and then blame the stone for failing. A crystal is a tool that can help you, but like any tool, it cannot produce any results alone. It cannot do the work for you. You must interact with it and be open to its influence, listen to its "voice" and put your own energy and skill into the process.

For example, you decide that you are too timid and would like to be more confident and outgoing. You might charge your crystal to help you deal with this, to take your fear away and help you become more confident. Or you might want to program the crystal to help clarify and strengthen you while meditation and thus find the cause of your fear so that you can come to terms with it. Either approach would work, depending on your individual nature, but you must put yourself into the process. In this case you must be willing to release your fear and, in situations where you would have felt self-conscious and shy, make an extra effort to consciously focus on the crystal's energy, feeling it send the strength and confidence to face the situation comfortably and with grace.

If you have suffered a loss, you might ask help in finding joy in the life around you, to turn the inner pain into the energy to reach out and heal. In this case you must make an effort on your own to reach out beyond your pain and reconstruct your life in a new and positive way. If you must be around negative, harmful people, you can ask your stone to receive the negative energy and charge it to provide protective, strengthening, healing energy for you, or to deflect it and send it back to the universal pool of energy rather than come into contact with it at all.

You must remember that this is an interactive process. To begin with, you must really allow the stone to have the problem you have programmed it to deal with. Let go of it and let the stone help you. Do not continue to generate the feeling or condition as fast as the stone takes it away. It helps if you have programmed it to change the energy and give you back something positive. Nature hates a vacuum, and you will automatically and unconsciously seek to fill the spot left with something else when the old negative pattern leaves. You must consciously put something positive in its place this is part of accepting the help that the crystal gives you.

Stones are very simple and, as with any relationship, you will get back what you put into them. In this case, you must participate in the energy-exchange cycle. You must allow the stone to take the vibrational energy of a certain frequency and pattern (your negative feelings, for example) and then accept this energy in its new more positive and constructive form after the crystal restructures and amplifies these energies according to your instructions. But you must allow this cycle to take place. You must be willing to accept the positive energies and make them part of you by using them and acting on them. You might also occasionally "love" at it, or say thank you to it, consciously acknowledge and appreciate what it is doing, while putting some positive strength into it.

WEARING YOUR STONE

There are many things you can do with charged and programmed stones in order to make their energies a part of your daily life. You can have them in your window sill or on your desk or dresser to work with the energies in a room. You can put them in plants so that their vibrant energies interact with the green living things. You may keep a selection of them in a container near at hand in case you want to take one out and feel its special quality of energy. A personal crystal that you have cleared and charged with a specific purpose in mind is generally worn someplace on your body so that its particular charged energy can directly influence you as you go about your life.

Once you have cleared and charged your crystal, you

should carry it with you somewhere on your body. In order for the stone to function properly, it should be in line with the main axis of your body's energy field and, ideally, somewhere near the center of that field - that is, somewhere over your heart. This is where it will have its maximum beneficial effect. It will have a difficult time interacting with you if leave it in your purse, briefcase, dresser drawer, or window sill. Please do not keep it in your pocket with your keys and change; this will chip and scar it badly. One exception to this practice is that of wrapping a crystal carefully in leather, fabric or metal, so that it is not battered by the loose objects. This can be a fun and interesting way of not loosing your keys by programming the crystal to always let you know where they are.

Ideally, it should be hung vertically, that is, with its center axis running up and down parallel to your spine, between your throat and the solar plexus. Many people wear their crystals in a soft bag made of leather or any natural fiber such as silk or cotton and hung around their neck on a soft cord. Others prefer a metal cage on a chain that encloses the stone. This will be fine as long as the stone can hang vertically and does not bounce and rattle around. The most successful way we have found to wear crystals is hung from a metal cap on a chain. The cap serves not only the function of keeping the stone in alignment with the body but also of providing it with an effective and uniform energy base as well.

Crystals grow from a seed that has attached itself on a base rock. This base rock provides a natural grounding and uniform energy intake for the crystals to draw energy from, like the earth and root structure beneath a plant. The energy field of a crystal moves from its base and emits a signal from its termination in a focused straight flow creating a signal similar to a radio signal or note of music. When a crystal is removed from this base rock, it is deprived of this natural energy base. The flow is interrupted, and the intake energy can become unstable and erratic through the fractured and disconnected base causing the signal to be less than sharp and clear. By cutting the base of the crystal flat, we establish an even field of intake, and by putting a metal cap across this cut end, we provide it once again with a coherent energy base from which to draw, giving it a regulated field of intake. The cap also serves the function of "qualifying" or "coloring" the nature of these intake energies according to the energy properties of the type of metal used. Silver is very effective for this because it provides a smooth, uninterrupted flow of energy and has the least specific influence of the available metals. It leaves the stone free to emit its clearest signature energy signal in the way that is most generally harmonious to the wearer.

We do not recommend using crystals that have had their caps electroplated onto them. The strength and nature of the electrical currents involved in the plating process often seriously alters and warp the internal axis vibration of the crystal, leaving it weak, erratic and permanently damaged.

There are many other variations on the placing of crystals within the energy field of the body. Some healing techniques involve placing a variety of stones over effected areas of the body's energy field. This is thought to not only address the area that is directly malfunctioning but also to bring it into balance with the other less affected areas of the energy field. The premise is using the combined energies of the crystals to balance and redistribute the energy into a harmonious healing field. Occasionally when a part of the body has been injured and the healing is likely to take an extended period of time - such as in the case of surgery or a broken bone - a specially charged crystal can be worn attached directly to the affected area to strengthen the body's auras and, as a result, the healing energies in that area. Doubtless as you continue to work with crystals and stones you will discover and develop a variety of techniques that work well for you. We encourage you to experiment and investigate. While you do, keeping a journal will be an invaluable aid in observing the long-term effects of your crystal work and can help to reveal subtleties that would go unnoticed over an extended length of time.

PRIVACY

People often wonder whether or not it is advisable to allow other people to handle their crystals. Opinions on this vary widely. Generally speaking, once you have cleared and charged your stone it should not be harmed by the casual admirer picking it up. In general social circumstances, the worst that could happen might be that the stone would pick up some envy along with the admiration. As you become more experienced, you will find that if you have cleared and charged the crystal adequately, it will not usually be affected by other people unless you allow it to be. However, when you are just beginning to learn how to work with them, a good rule of thumb is: Do not do what you are not comfortable doing. If you would rather not have people touching your crystal, it is perfectly acceptable to ask them not to do so. In fact, it is also a good practice to remember to ask permission before picking up a crystal that does not belong to you. Some people do not mind; others do. You will never know unless you ask first.

WHEN THE WORK IS FINISHED

Having your own personal crystal does not mean that you will wear it every minute of every day for the rest of your life. There should come a point in your work with it that the goal has been accomplished. You may, however, not realize this on a conscious level. One good clue that your stone has done its job is, when you take it off to bathe or sleep, you suddenly feel pounds lighter to be relieved of its weight. Another clue you might notice is that you just forget to put it on and don't miss its presence for hours. If you are unsure about how you feel, try meditating on it. Examine what you are feeling and whether or not your goals have changed in the course of your work. If you feel that your work has not been completed or the goals not achieved in the manner you desire, you might wish to cleanse and recharge it. This could particularly be the case if , in the course of your work with it, you have had some particular personal revelations or changes that change your personal focus and outlook. Reprogramming a stone to align better with these new realizations can give clearer and more harmonious focus to the changing nature of your work. Or, you might consider choosing a different stone to give that one a rest. If this is the case, you might cleanse it first, then hang it someplace quiet in the sunlight to revitalize itself.

INSULATION AND STORAGE

When you do not wish to wear your stone, you should put it in a safe place where it will be physically protected from accidental knocks and shocks, and from random energy vibrations as well. This is especially true if you wish it to retain a certain specific charge. Wrapping the crystal in silk is an excellent way to protect it on both levels; it not only cushions, it insulates against static charge as well. If you also place them wrapped in a wooden box - such as oak or cedar - all your stones should be safe and insulated from both the world outside and each other, and thus will retain their charge almost indefinitely. A plastic box will work to insulate your piece almost as well as a wooden one. A stone box is generally not recommended because the stone of the box may well have a charge of its own that you do not wish transferred to the contents.

In the future course of your work you will probably want to experiment with more than one crystal to find out what the various types of crystals can do for you. You may also wish to have different stones charged for different functions rather than letting one do the work of many. This is definitely a subject you should track in your journal. You may include a description of your crystal work with your regular entries, or you may want to set up a separate section devoted to crystals and stones.

CHAPTER 5
THE UNIVERSAL SPECTRUM AND COLORED STONES

THE UNIVERSAL SPECTRUM

When white light passes through a prism it separates into the seven basic colors. When spirit enters this universe, it expresses itself as separate elements that are sometimes called the Seven Principal Rays. There we can see the Law of Correspondences demonstrated. The Seven Principal Rays are the essential forces of this Universe and are to this universe of matter and form as the chakras are to your body. Each expresses a different frequency or type of energy, and you can explore the qualities of these rays of energy by using the analogy of color and chakras. Colored crystals, metals, physical systems, individual personalities, and much more correspond to these archetypal universal energy expressions.

Color can be described as a function of wave length and frequency; this means vibration. In other words, color is a form of energy - one that you perceive with your eyes but that exists and affects you on many other levels besides the visual. Pure light has no color, existing as unqualified radiant energy. When it encounters and passes through a prism, it separates into its individual components, becoming visible as the spectrum of visible light; that is, a rainbow of seven basic colors.

Because the vibrations of color energy exist in more than just the visible range, you are affected by them on more than one level. A specific color energy will affect you on all levels simultaneously, each in its own particular way. You should keep this in mind when you are working with a specific energy. The qualities and characteristics of the color you are using will affect all of the levels of awareness in that particular range. Below is a chart demonstrating how these energies manifest and correspond to the different levels of awareness.

LEVELS OF AWARENESS, COLORS, AND CHAKRAS

COLOR	CHAKRA	PHYSICAL	MENTAL	EMOTIONAL	SPIRITUAL
Red	Base	Instinct	Necessity	Anger	Passion
Orange	Belly	Action	Evaluation	Stability	Balance
Yellow	Solar Plexus	Adaptability	Empathy	Intuition	Transformation
Green	Heart	Growth Healing	Understanding	Love Compassion	Synthesis
Blue	Throat	Expression	Interpretation	Awareness	Control
Violet	Brow	Thought	Focus	Interest	Discrimination
Purple	Crown	Energy	Clarity	Serenity	Higher Self

Clear light represents energy without specific form that embodies and emits all vibrations without qualification, reservation, or retention. It is the purest form of energy because it contains the potential of all colors centered within itself. This energy can then be separated with a prism into all the spectrum of colors. Clear is unqualified, unbounded, and unlimited. It is essence without form and potential without boundaries.

The red ray is power in its most primitive form, manifested in the Base Chakra where the primal earth energy first enters the body at the base of the spine. Red energy is passionate and energetic. The red ray contains lust and anger as well as survival and determination, and although it is dangerous in some aspects, without it there would be no survival on this physical plane.

This color is an aspect of the red energy also coinciding with the Base Chakra, but linked to the heart chakra as well. It is the resonance of passion on the physical plane and love on the spiritual and emotional plane. This is the love that bonds us together and flows from the mother to nurture and nourish the child. Rose is the energy that enables the individual not just to survive but to grow and flourish.

84

The orange ray banishes the darkness. This is the energy of hearth and home, the love of the young, the safety of the dwelling. It is the ray of true courage that defends the just cause. Also, orange is the ray of self-awareness and pride in personal achievement.

Yellow is the color of sunshine and joy - the heart of dancing and song. The yellow ray is the Power of Hope expressed as the Power of Mind. This influence often leads to change because it is manifested as brightness and cleverness - the joy of the spirit in all things that are new and unique. This is curiosity in the most positive sense, indicating all the capacities of the mind and spirit that see the world as new and fresh.

Green is the color of regeneration, restoration and renewal on both the spiritual and physical levels. It is also referred to as the color of the heart because the heart is the metaphor for the ever-springing force of love and renewal. This is the ray of the Mother and of Mother Earth for it connotes the ever-renewing bounty of the physical world. This is the ray of selfless nurturing and creativity without possessiveness. As such it is the color of artists and gardeners because they participate in the ongoing unfolding of the beauty of the world, while releasing their works to become parts of the sum total of its richness..

Blue is the color of pure will and intention, the clear focused force of perception and expression. Blue helps in bringing down the barriers to self-awareness by bringing the nature of these barriers into true perspective. Blue also breaks down the barriers to true sight on all the planes by opening the inner channels, focusing and refining the energies, allowing conscious awareness of the higher vibrations.

Violet is the ray of understanding those matters that extend from and deal with the Higher Self and its relation to the larger worlds - sometimes called the Multiverse. This is the path by which the voice of the inner self reaches the conscious self. Its center is the brow. Violet energy is also connected to the Higher Consciousness energy of this Universe and can be a great guide to wisdom of many kinds. Violet is a powerful path to healing because it reveals the higher nature of the causes of many situations and conditions.

Purple - royal red violet - is the color of power and balance because it is the perfect balanced joining of red, the primal animal drive, and blue, that is the pure expression of evolved higher will. The Purple ray centers at the crown of the head and is sometimes referred to as the crown lotus of awareness. It is both passive and active, centered between the white and black, and all the others will answer to it. This is the color of the true Ruler and Monarch and is sometimes ascribed to the Crown Chakra rather than Clear or White because it holds the understanding of the complete nature of the human being and from this understanding come both discipline and compassion.

BEYOND THE RAINBOW

We have covered clear light and the spectrum that appears when clear light is separated into its component colors. This refers to visible light, but there are colors that are not part of that spectrum. Instead they have energy potential because they include a wider range of its potential than just a single part. These colors are white and black. These are important working energies that exist in what might be called a class by themselves.

White light contains the full spectrum of energies unified into one ray of single purpose. This is the centering of the spectrum and the pure focus of its concentrated energy. It draws all the individual rays into itself and knots them together, supplementing what is lacking in some with what may be in overabundance in others. This is because it represents that which is truly centered in itself and lacks nothing. It is enfolding and uncompromising. It dictates nothing but manifests as all in one.

Black draws all rays of energy into itself but emits nothing. It is the reservation of power because it contains everything and gives away nothing. By its nature, it balances and levels all vibrations and could very well be called a natural grounding element. It is the perfect mask, for it has no reflection and reveals nothing except its existence.

COLOR IN CRYSTALS AND OTHER STONES

Using a clear quartz crystal brought the concepts of energy work very close to you and gave you an actual tool to touch and manipulate. In the same way, you can explore the energies of color in more depth by using colored stones and crystals.

Color is caused by an extra atom or atoms being added to or trapped within the molecular structure of the crystal. This changes the internal lattice structure of the crystal slightly causing it to reflect and refract light differently. So, a clear crystal with the addition of extra atoms to its molecular structure becomes a colored one. This also changes the frequency of its vibration and, therefore, the way in which it reacts when you work with it. Color specifies and qualifies the energy properties of a crystal or stone. It can also be caused when molecules of a different type are trapped into the basic structure making what are called inclusions. Sometimes inclusions appear as flecks, clumps, or striations and can be seen by the naked eye. Sometimes they are evenly distributed, sometimes randomly. Inclusions also change the crystal's energy properties.

You may wish to experiment with colored crystals to find out how they act and react, but the cost of some of the more unusual ones may be more than you feel you can afford. A workable alternative to perfect colored crystals is tumbled stones. Nearly every rock and mineral shop will have a bowl or basket full of brightly colored shiny pebbles. They usually come in a wide selection of colors and types and are reasonably priced. They will not have the clarity or intensity of whole crystals with perfect terminations, but the molecular structure is the same as it would be in the complete crystal. Therefore, their vibration frequency will be the same although not as strong. You may want to carry a small bag of them with you to be handy when you feel the need for a particular type of energy.

Since color is a form of energy, the presence of color adds its own dimension to the energy characteristics of any crystal. For instance, the term "quartz" is usually used to refer to the clear or milky variety, but the quartz family is a large one that also includes amethyst (that is, violet) and citrine (that is, orange). Either of these two will react in considerably different ways from the clear variety. The type of energy in each colored crystal or stone corresponds generally to the chakra of the same color. Stones and crystals of these colors can be used to supplement and/or augment the chakra energies and can be of invaluable help in channeling and using these specific frequencies.

CRYSTAL FAMILIES

Just as all animals are not dogs, not all crystals are quartz. The word "crystal" refers to a type of mineral formation, and many minerals form in the earth in this way. Each crystal family is characterized by a unique shape of termination and number of sides. As you might expect, this gives it a unique energy quality as well because the configuration of the energy lattice is different and distinct from the other families. There are often many different colors represented within each of the families. Tourmalines come in every color of the rainbow. Members of the Beryl family include both aquamarine and emerald. Although Sapphires are generally thought of as blue, they come in many shades, from orange to white to purple. In every case, it is not the color of a crystal that gives it its identity but its formation. To take this one step further, when color is added to their unique vibration signatures, these crystal families contain members that have the closest possible resonance to a particular archetype color in the spectrum. For instance, an emerald is the green member of the Beryl family, and it embodies the purest form of green energy, closest to the pure harmonic. This means that you can experience the unique and specific energies of color through the crystal that most closely exhibits them.

Color Changes

As you begin to work with a colored crystal you will probably notice some changes occurring in it. When you clear and charge it, as you run your own special current of energy through it, the atomic particles will begin to line up even straighter and align with each other. This allows light to be transmitted through it more easily and will cause the crystal to appear clearer, less cloudy, or lighter in color. In some crystals, this color shift can be seen dramatically - the different colored "fire" in opals is an excellent example. As they are worn their colors become much brighter and will tend to exhibit different colors depending upon the mood of the wearer. Amethyst and some tourmalines will become noticeably clearer and lighter in color. You should not be alarmed when this happens. It is an indication that your crystal is working just as it should in harmony with your changes and personal vibrations. The only time for concern is when the color in a crystal becomes grey or dull. This is an indication that it needs to be put aside for a while to restore its natural energy. Don't worry, it is not damaged, it just needs to rest.

COLOR DENSITY AND SATURATION

The power of the color is in direct relationship to its degree of saturation. This means that the more atoms are added to the basic structure to make a particular color, the more your crystal will behave in the manner of the color. You may or may not find this desirable. A garnet may be a lovely shade of rose, red, or purple with the coloring agent in balance with the basic atomic structure of the crystal so that each contributes to the other and makes one harmonic unit that can be easily worked with. However, when the coloring agent appears in such abundance that the crystal is so muddied as to appear black, it may lose some of its usefulness because it does not resonate clearly with the signature pattern of the crystal type because of the large amount of the coloring agent. You should also remember that although a stone looks black, that doesn't mean that it is. We are working with atomic elements here, not strictly optical properties. A crystal can have such a dense saturation of colored elements in its structure that light will not pass through it and this will make it appear black. You should always use any stone according to the coloring agent rather than how it might appear to the naked eye.

OPAQUE STONES

Of course there are many other different types of stones than crystals. Their colors and color combinations are seemingly limitless, and each has a special subtle property that is unique to its kind. It would be impossible to discuss or even list them all here. When approaching these stones, you can begin to evaluate their potential properties by judging them according to their color and family type if you know it. If you find one that takes your fancy and won't leave your hand but have no idea what it is or what it's good for, here's a way of approaching it from the intuitive side.

While in a meditative state, hold it gently in your hand, much in the same way that you worked with your quartz crystal. This does not have to be a deep trance state. It can be a lightly relaxed and receptive state of mind, like that state when you begin your deep breathing but before you go into a deeper meditation. In this receptive state of awareness you can even have a kind of conversation with it. Before you clean and charge it, ask it what it does and why you felt that you needed its energies in your life at this time. Allow an answer to come to you; then later, when you have finished your meditation, write down the answer you received in a notebook on its own separate page. As you work with the stone, be aware of how it makes you feel and what differences you feel or sense in yourself and your life, and enter these observations in your notebook also.

This process will serve two purposes. It will encourage you to be aware of your stone's activity as well as develop your subtle sense of intuition and inner observation. Over time, it will also result in a notebook that will become a valuable resource of information and observation.

Clear crystals carry the key signature pattern of the family of crystals they represent. They are the pure essence of that family because they have no color energy added to change the true note of their energy lattice vibration. These stones carry the full potential of the family they represent because they are not limited to any one ray of energy. They carry a full range of potential energy qualified only by the basic resonant nature of the family type. They can be used by themselves or be combined with colored stones to focus and amplify the energy they represent. They can be worn by those who are experienced at working with crystals because they carry the purest singular energy of a crystal type or family. They can also be worn by the beginner because they are not too specific and direct but permit a wide latitude of working ranges. Clear quartz is the primary crystal for this energy signature.

Ruby and garnet are first among the red stones that help to harness and channel this energy. Red stones may be worn to strengthen the will to live and bolster up the courage or the ego. They can provide a tap not only to the vitality of the body, but also of the spirit. This is not merely the energy of endurance, but of victory.

Rose stones include rubelite tourmaline, rose beryl (morganite) and rose quartz, as well as some rubies and garnets. Any of these stones can serve to heal and reinforce the primal will to live, and can strengthen and revitalize the life forces by reinforcing the individual's connection with the physical universal source of energy. They attune to the vibration of spiritual and emotional love and physical passion. The rose vibration resonates in harmony with the heart chakra and makes these stones exellent for healing energy and the ongoing rejuvination and restoration of all the energies for all levels of awareness.

Citrine, carnelian, and Pad Parajah (orange sapphire) are the most prominent of the orange stones that stimulate courage, pride, and determined identity and self-worth. They may be used to hearten a warrior or comfort a child's fear of the dark, or help with shyness and stage fright.

Yellow stones such as heliodore (yellow beryl) and topaz stimulate the wit and intelligence. Their nature is the quickness of laughter and the freedom of exploration that manifests as pure mental energy. They can also help the wearer to overcome both fear and fatigue, and stimulate hope and optimism.

Green stones are those of the healer because not only do they contain the healing and restorative energies, they also restore the healer. They constantly revitalize the systems even as he or she works with the healing skills. Not only do they help the heart chakra; they also aid in balancing the flow of energies in the horizontal axis. Green stones can be worn above any chakra of the body, and they will strengthen that area according to the energy of that center. They are often worn with rose-colored stones to reinforce the life force energy. Emerald, peridot, and tourmaline all act as restorative and strengthening stones.

Sapphire, tourmaline, iolite, and lapis lazuli are the stones to open the gateways by which the focused will is manifested on the physical plane. They will help the wearer to reach outward to access information and experience available on many levels while retaining individuality and personal integrity. Lapis lazuli is particularly effective. It is sometimes referred to as the Seer's Stone because of its great ability to open the channels of sight and perception.

Amethyst is the purest nature of the violet stones, although some tourmalines will have a pure violet color. Its power lies in its potential to link together the rays of knowledge and make the individual consciously aware of the nature of the ills. It is a great stone of strength for the practitioner because it allows access to all forms of knowledge and energy from other realms along with the sum total of personal and collective knowledge. Violet stones are also effective in helping the user access the higher self.

One excellent example of a truly purple crystal is the siberite tourmaline. Others are certain rubies and garnets. Purple stones can help in understanding complex natures. They should not be used to seek or enforce a balance; rather their function is to assist the already balanced individual to work within the balance to harness all energies wisely and well.

BEYOND THE RAINBOW OF STONES

White, black, and brown stones have their own unique properties, just as white and black have unique qualities beyond the range of the spectrum. They work with the energy system as a whole rather than individual resonances or frequencies of it.

White stones draw all energies into themselves, then unify and return them. They act as natural buffers, creating a reservoir of energies from which the flow is mended and balanced. White stones are generally those that have structural patterns that cause them to appear milky - such as white jade, alabaster, milky quartz, and moonstone. In choosing a white stone, care should be taken that this milkiness is regular and smooth.

White stones can, in some cases, be used in place of grounding stones, although they do not specifically ground out energy. True grounding stones remove excess or extraneous energy; white stones repattern the energy formation in order to complete it. The net result is the same, though, because they have a soothing and calmative effect as well as a focusing one.

Black and brown stones are absorbent in nature. They draw in and purify energy, reducing it to its lowest possible harmonic frequency by removing all static and wave form anomalies. In this way they literally edit out all excess static from the waveform while giving back one smooth curve of energy. This works extremely well to take the edge out of nervousness and hyperness without making the wearer "dopey or spacey". They will not add power or energy, but they will help the wearer clarify his or her own process by removing what is not necessary. They will also help a great deal in grounding after work has been done, and allow the individual to rest fully. Black and brown stones can also be kept with other stones to ground them naturally and to keep them from picking up unnecessary stray energies, or to purify and ground them when their function is accomplished. Black, and to some degree brown, are self-clearing by their own nature. The more they draw, the stronger and more stable they become. Once set, they are virtually impossible to overload.

Black tourmaline is the strongest of the grounding stones, followed at a close second by black sapphire. Black jade works especially well on women. Hematite (and any other metallic-based stones such as pyrite) is NOT a grounding stone. Hematite does often have a calming effect however, the red iron energy in the hematite strengthens and unifies the system in a balanced and harmonious way. Often stress and tension drain the resources on all four levels of awareness causing increased agitation along with diminished ability to function. By acting as an energy restorative it helps the individual to adapt and cope, giving the effect of grounding and calmative. Its basic component is iron ore and will behave in the way of that metal rather than as a black stone. Smoky quartz is a very gentle calmative that, because of its quartz nature, will also have a tendency to focus and amplify.

METALS

Metals have qualities similar to crystals. This is because, when metals are worked and drawn, their structure "lines up" in an extremely tight and regular way, allowing it to easily transmit energy of all sorts. This property is called conductivity and works just as well for subtle forms of energy as it does for electricity. When you are choosing or working with a stone, you should take into account the properties of the metal that holds it. This does not mean that you can only have the stone set in its corresponding metal. It means that you should consider the way in which you want a stone's energy to be broadcast or made available for your use. Some metallic energies are very focused and piercing; some are gentle and permeating. Some, such as iron, are generally not suitable to be worked into a setting for a stone or are not available.

Iron and steel correspond to the color red. Steel is the only metal that when worked will reach a truly crystalline structure and, because of this, will draw and cycle energy of its own nature in the same way crystals do. Steel represents the nature of war and command, and can be used to access the primal life energies of the physical plane. Its nature could be described as both warlike and passionate. Iron and steel are known as the metal of Mars. It is rare to find a stone set directly into steel because of the difficulty in working with so

CENTER	COLOR	CRYSTAL	METAL	CALLING
Crown	Clear	Quartz	Platinum	Angel
		Diamond		Higher Self
	Purple	Tourmaline		
		(Siberite)		
Brow	Violet	Amethyst	Electrum	Monarch
Throat	Blue	Sapphire	Silver	Seer/Teacher
Heart	Green	Emerald	Copper	Healer
Solar Plexus	Yellow	Topaz	Gold	Dancer
Belly	Orange	Citrine	Bronze	Warrior
Base	Red	Ruby / Garnet	Iron	Artisan / Smith

hard a metal. Even when stones are set into the hilt and pommel of a blade, the hilt and pommel will be made of a more malleable substance such as wood or bronze. Iron and steel are very demanding energies to work with, and it is not recommended that you wear them in line with your vertical axis for any length of time such as general daily wear.

Bronze is an alloy of copper, tin, and sometimes zinc. It corresponds to the color orange and is highly compatible with the orange stones. It is said that bronze gives additional physical strength because it stimulates the flow of energies that support the body's external physical structure in the muscles and circulation. It stimulates courage and joyful warrior passion. Bronze will not conduct as well as steel, but it will take a specific charge and hold it clearly.

Gold is sometimes called the metal of dominion because of its ability to drive and focus any charge with great clarity. Gold's nature is active, and it will magnify and enhance the qualities of any stone set in it. It is also referred to as the metal of the Sun- both because of its color and because it operates throughout the energy system. Especially when worn over the heart, gold will extend the influence of whatever stone is set in it throughout the entire energy system just as the sun's light brings its life giving energy to virtually all aspects of life on Earth. It also acts to bring about a balance between all the energy points because it encourages even and harmonic flow of energy.

Copper has the energy of the union of rose and green rays and, as such, contains a balance of the energies of the heart and base chakras. It is an excellent conductor but does not retain a charge for long. If worn for a long time, it will tend to take on the nature of its wearer and become a strongly personal talisman. When it is worn or worked with any other metals, its nature tends to add a mellowing influence and flexibility to otherwise harsh or incompatible or "brittle" vibrations. Astrologically, copper is called the metal of Venus.

Silver is called the metal of the Moon because of its subtle and flowing nature. It has a subtle, peaceful, and passive nature and will not "color" or interfere with the nature of the stones that are set in it. It will help distribute the influence of energies throughout the system and to make those energies easily acceptable to the system as a whole.

Electrum is an alloy of gold and silver in equal proportion. In ancient times these metals were occasionally found together in nature. Its nature has been highly valued because it possesses the ideal qualities of both gold and silver. It can be active as well as passive and able to send and to draw. It is analogous to the color purple/violet.

Platinum is the most potent of the noble metals but is extremely difficult to work because of the high temperatures involved and because of its extreme hardness. It is a brilliantly forceful driver metal and will amplify and drive the influence of any stone it holds. It can be used with great effect and has the added advantage that its nature is not "automatic" because it will not function unless the wearer intentionally causes it to do so. This means that even though it is very potent and strong, it is unlikely to inadvertently overdrive the wearer.

Future Study and Reference

The study of how stones and metals interact with each other and with the chakra system is an intricate and fascinating one. The theory and practice of the making of complex balanced talismans of stones and metals could fill a volume in itself. At the beginning of your research, it will be important to you to keep a record of how a stone and its surrounding metal setting (if any) affects you, what you use them for, and how they look and behave when you do. This will build up a working knowledge base to which you can return time and again. Do not be surprised if your preference and methods change as you grow and develop. It is only natural that as your sensitivity and work progress, your energies will change as well and with that change you will want to use different stones and combinations. You may want to include this information in your journal, or you may want a separate notebook or card file to accommodate the flexible nature of the information.

CHAPTER 6
HEALING ENERGIES: POWER GAMES AND FIRST AID

Crystal work is only one of many things that you can do with the horizontal power axis of your body's energy system. Your arms and hands are your natural expression points of your chakra system, a natural tap on the universal flow of energy as it passes through you. This is the first way in which you will learn to express your energy in terms of power to influence the world around you.

But, we need to take a moment out here and talk about the "P" word - POWER. Most people are afraid of power - afraid of what will happen to them if someone else has more than they do, or afraid of what they'll do if they have too much. Power is a very scary word and, in this case, conjures up images of evil magicians throwing balls of green fire at each other. The real truth is that power is neither bad nor good - it is just like the electricity that rests quietly in the wall sockets of your home. You have the choice of lighting your home and cooking your meals with it or you can stick your finger in the socket and electrocute yourself. It's entirely up to you, but the electricity itself is neither bad nor good, it is just an elemental force. Since you were a small child you have probably been taught how to use the power in the socket properly and not to get hurt in the process so that by this time you are probably pretty relaxed with it.

In the same way, you have also been taught how to be responsible for what your body does as it grows - how not to strangle the cat when you hug it just because you're bigger than it is, how not to break the dishes and trip over the furniture. Having a strong body is another kind of power that is neither good nor bad all by itself - it depends upon how you use it. When you begin working with energy, what you are doing is really no more scary than getting your body in shape by lifting weights and doing exercise. The power we are talking about is the natural fabric of the Universe. Every human being has it whether they are aware of it or not, and it also exists in every living being in Creation. You are just learning how to become aware of it and make it work for you. It is still up to you to choose how you will use it. And like having a strong body, you need to practice some to get comfortable with what you can do.

THE HORIZONTAL AXIS OF POWER

The center of power for the horizontal axis is in the Heart Center where the energy currents flowing through the body reach their most balanced point. People generally speak of this center as being either green, rose, or white, depending upon their school of thought or the way in which they perceive it. The horizontal axis runs from the heart across the shoulders, down the arms, and out the hands. The left or recessive hand is the negative pole - it is the intake receiver. The right or dominant hand - the positive pole - is the outflow terminal. This is true if you are right handed; if you are left-handed, it is reversed.

Now put your palms together and bring your elbows slightly up. Envision the energy of your heart flowing in a circle around and around through your heart, out your right hand and into your left. Imagine that with each circuit the energy gets stronger and stronger, amplifying itself and drawing more current from the vertical axis. Breathe deeply and feel the cycle accelerate and its strength increase. When you are comfortable with this and can feel the energy flowing freely and easily around and around, keeping your palms flat, move your hands about an inch or two apart. You will feel the energy "stretch" between your palms. It will feel just slightly cool. You may even be able to see it as a vague distortion or slight shadow stretched like taffy between your hands. If you practice at this and really get it going, you may even be able to see a faint visible glow in a darkened room. Get a friend to put their hand between yours and ask them how it feels. Please remember that this is all perfectly natural and that this energy/power is there all the time. What you are doing here is just using a little feedback to get it pumped up some, that's all.

So now that you've got it going, what do you do with it? Good question. You have probably noticed that this is exactly the same procedure you used to clean and charge your crystal. At that time, you simply added a thought content onto the energy flow - you intended something to be. This technique is good for an amazing variety of things. Suppose someone has cut their finger in the kitchen and you want to stop the bleeding. A little bleeding will clean out the wound, but there isn't any reason to go on and on with it. Cover the wound with your dominant hand, look the person straight in the eye, and tell them that the bleeding has stopped and that the pain is going away. Then ask them if this is okay. (Getting their permission is very important.) When they have nodded yes, know and command the bleeding to stop. A slight surge of energy should do the trick.

Covering the wound is important. As a general rule, you will not be able to override someone's free will and belief that their finger is bleeding all over the place if you let them keep on looking at it and enforcing the belief in the injury.

Yes, it really works, and it is just that simple. We have used this on countless adults, children, and occasionally on small animals. With slight variations of intention you can expand this to also take care of bruises and swelling from injuries and insect stings. However, it is important to get to them quickly before pain and fear enforce belief in the wound. After that, you can still do something for them but it takes a lot more push. This is what we call band-aid magic and it is a wonderfully handy skill to have.

After you have stopped the bleeding and shock it is advisable to go ahead and put a physical bandage on it for two reasons. The first is very practical because you want the injury to stay clean and heal cleanly without opening up again. Secondly, it will keep the person's mind off it and reinforce belief in the healing of the area.

ETHICS

This is the time we need to talk about the ethics of being any kind of practitioner - and the moment you start using your talents on other living creatures you are a practitioner of sorts. Free will is an extremely important concept. If you had an exceptionally large strong body, it still would not be right to enforce your personal will and beliefs on someone, regardless of how convinced you were that what you believed was right and they were wrong. The only time this is permissible is in the case of snatching small children and animals out of the path of moving vehicles, keeping babies from ingesting poisons, dragging unconscious people out of the path of harm, etc. These individuals are not capable of making decisions for themselves, and the line of difference is pretty easy to distinguish. It is definitely NOT all right to help a person whether they want you to or not. This is a serious violation of their rights as an individual and would be just the same as holding them down and forcing aspirin down their throat if they did not want it. If a person has requested your help, then it is fine to help them, but then only to the extent that they have requested your help. If, in the course of fixing a simple thing, you find there is a larger and more serious problem, you should discuss it with the individual first and ask permission to deal with it. If a person says that it is not all right for you to help, that is their decision. They have every right to keep their injury, pain, or whatever if they want to, no matter how odd that may sound to you.

BACK RUBS

Between crouching at computer terminals and desks all day and after-work activities, many people are plagued with a wide variety of lumpy aches and pains in muscles and joints. There is a technique involving both hands that you can use on a friend which can really help alleviate a lot of their stiffness and soreness. This is a technique that you can combine with a simple back rub to handle the energy imbalance as well as the aches and pains. Pain is a form of energy. In the physical body, it is represented by tiny electrical impulses passing through the nerve channels. Like any form of energy activity, it affects the other energies it comes in contact with. It affects the individual on all levels of awareness, and in order for healing to take place, the energy of the pain must be dealt with. Any sort of pain presents a double problem. Not only is there the injury itself to deal with, but there is also the attending energy of pain and inflammation. Also, the body's energy field tends to withdraw and become thin around the area in order to minimize the awareness of discomfort. And so begins a cycle of discomfort and withdrawal - removing the very energies that the physical organism needs most in order for comfort to return and healing to be effective. Before you begin with this, however, you should make sure that you are well practiced in centering, grounding, and shielding so that as you are working, you do not unintentionally retain any energy you don't want.

Begin by building up a cycle of energy between your palms as we have just described. As you cycle the energy around and around, the dominant hand is the outflow terminal and the recessive hand is the intake. As you massage a sore muscle, envision and intend that the soreness - the negative dark energy - is drawn in to you through your recessive hand and up your arm. Remember, no matter what the specific nature is of the pain or soreness, it is only a qualified form of energy. Using your will and your intention, you should change the color or content of it (however you personally perceive it) into positive, white, strong, warm, healing energy and pass it through your dominant hand and into the sore muscle.

This is a very important technique and will stand you in good stead in many different areas. Changing the nature of energy saves a good deal of effort and potential grief later on. By using the energy generated in the injury, you will not be in any danger of overloading the area because you are using energy to the degree it is generated and no more. If you need more, it is always available, but this is also a way in which you will not become drained of your own strength in trying to help

105

someone.

Drawing out the negative energy will also help a great deal in getting the injury to heal itself. Drawing out the pain breaks the cycle of swelling and stiffness associated with sore muscles and allows you to work the flexibility back into them. It also creates a "vacuum" or empty spot that can much more easily be filled with the healing positive energy you are going to send into it. But once you've drawn up the negative energy remember IT'S NOT YOURS, AND YOU SHOULDN'T KEEP IT. Keep it moving. Change its content and PUT IT BACK WHERE IT CAME FROM. This is the big "secret" of how not to get fried by someone else's problem. It is just like picking up someone else's dirty laundry. You don't want to keep it forever just the way it is and call it yours. You wash it up and return it dried and folded. No problem.

THE CYCLE OF ENERGY

This demonstrates very clearly an important natural law of the universe. You have certainly heard it said, "Nature hates a vacuum." That's true on any level you can name. The Universe is made up of one coherent fabric. It is not just a random mush of bits and pieces. If you wish to take something out, you must put something in its place, or the empty place will seek to fill itself with whatever is most like the thing that last occupied it. If you wish to heal something by drawing energy away from an area, you must complete the healing by filling the gap with something else - love, healing energy, warmth - whatever is appropriate to the problem. By changing the content of the energy that the injury generated in the first place, you replace the undesirable energy with energy that is more desirable but that is enough like what you just removed to be compatible.

As the energy passes through you, you may also become aware of the nature of the injury. You may get pictures in your mind, impressions, or just have an instinctive knowledge of what you need to do. There are many different ways to receive and interpret the information that comes along with the negative energy. It is as though people have pictures of the injury "stuck" to the injured area. As you practice with this, remain relaxed and aware of what you are receiving and perceiving. It is not possible to tell you exactly what to look for because everyone perceives this information a little differently. Be assured that the energy that you draw from the area contains information about that area and the nature and extent of what the problem is. This is very useful information when you are looking for the right way to help.

HEADACHES

Headaches bring up a slightly different area of consideration. What do you do with the negative energy if you don't want to put it back where it came from, as in the cases of taking away a headache or in mellowing out someone who can't sleep? This is real energy and, if you keep it, it will affect you the same way it was affecting the person you took it from. You must always turn over or rework the energy before you release it. Change it to another "color" or change the content, however you perceive this change as appearing. If you do not change it before you release it, you will just be spreading nasty grey stuff around. Then, if you are not going to give it back, release it to the universal pool of energy. This is the general store of life force within the fabric of the Universe.

WEAKNESS AND TEARS: DRAWING THE EARTH'S ENERGY

You can take this technique another step forward in dealing with extreme emotional states such as grief and anger. It is necessary to an individual's health and well being to be able to express emotions. However, there are times when emotions can run away with a person and be every bit as harmful to them as an illness or injury. At such a time, you may reach out and draw away excess emotional energy in very much the same way that you dealt with physical pain energy. Emotional states tend to distribute themselves throughout the entire system rather than being limited to a single area although they are often focused through the head or heart. In dealing with this you should be advised that you must deal with it only as an extreme excess of raw energy. You should not become involved in the subject matter of it. Once the individual is calmed down, the causes of the upset can be dealt with. Take the energy away, and replace it with love or healing restorative energy.

You may also want to try just the opposite in the case of illness or extreme fatigue, and contribute energy to an individual to give them extra strength to heal and rest on their own. This should never be regarded as a substitute for rest, healing, or a good meal and should be used carefully. In this case you should not draw what you wish to give the individual from you own body's store of energy. This would only debilitate you. Trading one "basket case" for another is not what we have in mind here. Instead, draw what you need from the earth energy beneath your feet. This may take some practice, but is essentially the same action as cycling the energy through your hands.

Feel the earth beneath your feet. Now feel the power within the earth. Feel the warmth of that power resonating deep within the earth's core. Now you can feel it rising upward, warm and strong, pulsing through the earth into the soles of your feet. It flows through your legs and centers at the belly chakra, enhancing your whole energy flow. You feel strong and powerful with it. This is closest to the vital primal energy of survival and renewal that is lacking from the person needing your help. When you feel comfortable with the energy you have drawn up, bring it into your heart and let it flow into the individual you wish to help. Channel it through their heart so that it will be balanced and distributed throughout their system. Give them only as much as their system will accept easily. You will feel a resistance in your own system when you have given all they need. Then center and ground out whatever energy is remaining, and rebalance your own system.

GROUNDING

Everyone has a natural balance and harmony of energy levels that they maintain. This changes from time to time depending on your relative health and state of mind, but mostly you maintain an equilibrium. When you work with higher level energies, there will be times that the energies you are dealing with get "a little too much", when you retain more of their intensity and/or content than you wish to. For example, this can easily happen when you have found it necessary to draw on an extra amount of energy to accomplish a specific purpose such as a healing. When the need is over and the task accomplished, it is time to put this extra energy away and let it drain out of you so that your natural energy level and balance can be restored. Or it might mean that you have needed to draw away a great deal of pain or grief energy from a hurt or frightened person. Sometimes you can just get overwhelmed by being in an extremely energetic environment, and after you leave you would like to return to your own natural levels. What you need to do is "ground out".

Grounding literally means to let the excess energy flow back into the base from which it came. There is a pool of Universal energy that exists all the time. It is from this endless source that you draw your own energies and any extra that you need to accomplish a task. So when you are done with it, you can just let it flow back.

GROUNDING EXERCISE

In a waking conscious state, take a quiet moment. Take a deep breath and as you exhale, imagine that the excess energy is flowing downward through your feet and into the ground - all the way into the center of the glowing heart of the Earth. It is important to flow the excess energy all the way down because you do not wish to have it just soaking into the carpet or hanging around in your basement. It should go all the way back to its source.

If you prefer, imagine a pure white star as the source of all Universal energies. As you exhale, send the excess energies as a beam of light back to the star to be purified and reabsorbed. Either way you choose, be sure you send the energy all the way back. Then, take another deep breath and feel your energies come back into balance. Take a brief moment to reestablish your own comfort levels and center.

Centering, shielding and grounding are not techniques to be held in reserve for when you think you might need them, nor is it meant for you to develop an attitude of isolation or paranoia. The more you work with energies, the more sensitive you will become to them. You will not always have the luxury of quiet time and space to "put on" your shielding or "find" your center. Instead, these are qualities and attributes that you should develop as an integral part of your personal energy surroundings - a part that flows with you at all times. You should incorporate these images into your daily routines.

In general, it is absolutely necessary to remember about grounding and centering when you begin to work with other people. Accept only what is yours, deal with what it's necessary to deal with, but let the rest flow away. By remembering this you can be of real help to others, and your skills will increase enormously as you practice.

AURAS AND ENERGIES

Now that you have gotten a little experience with healing, we should begin to talk about energy fields in a little more depth. It is generally agreed that the aura is a haze or halo of light surrounding all or part of the body that, by its color, shape or pattern, can indicate the physical, emotional and/or spiritual condition of that body and the spiritual being that inhabits it. If you think that is a pretty vague definition, you're right, and you're certainly not alone in thinking so. Unfortunately, this is where the general agreement stops, and theories and opinions branch off in all directions. So, we will try to give you some general guidelines and explanations of what people see when they see auras, so that when you begin to work with this a little more, you will have some idea of what you're dealing with.

Part of the problem is that not all people perceive and visualize energies the same way. Some people can see energy fields visually; others feel them as temperatures or textures. To some they appear as colors, to others simply as bright, dull, or dark spots. It will take practice to develop your skill at discriminating what you see and then being able to change your focus and see differently to perceive different states of energy. But the primary difficulty is that the term aura is used as a very broad term to cover the several different levels and layers of energy manifestation that we have been calling the levels of awareness. This can get confusing, to say the least, until you get some experience with it and can begin to determine for yourself how you see and what you are seeing. Auras are the visible (or perceptible) manifestations of energy fluctuations in the "subtle bodies". The complicated part of it is that they are nested one in another, influencing one another's energy. It can be difficult to decode and unravel just what you are perceiving because they appear as though they are all manifesting at once. It takes some practice, some intention, and just some relaxing with it to observe how you see things before auras really start to give meaningful information. This is a key issue. You must become accustomed to how your own senses relay message to you. Some people see auras as colors, others as hazy intensities of light, others feel them through the palms of their hands, others say they "just know". It will take some practice and a friend or two who is willing to act as your guinea pig, to give you some feedback.

PHYSICAL HEALTH

The body is surrounded by a field of energy that it creates as a result of being a living organism. Its nature is basically piezoelectric; that is, chemically generated electricity like a battery. This is caused by the action of neurons relaying messages both to the brain and among themselves and the various systems of your body. This field is intimately interwoven and linked into what we could call the life net; that is the physical manifestation of primal life energy in this Universe. The combination of the body's electrically generated field with the primal life energy appears as the physical aura. All life forms have this physical energy aura, including animals and plants. It usually appears as a gauzy haze or envelope around the human body, usually six to eight inches wide. In plants and animals this varies.

When an organism gets injured, there is an initial burst of energy as the message of that injury is carried to the brain that stimulates the "damage control" and healing mechanisms of the body. (In the case of organisms with no brain like plants, the message is communicated throughout the energy field in general.) If the damage is extensive, however, and/or of long duration, the healing mechanisms can cause a general drain on the body's reserve of vital energy. This will cause the aura to appear smaller and less bright. As the drain continues, the being will withdraw from the affected area - a sort of spiritual amputation to relieve both the continued pain and the energy depletion for the rest of the body. If this continues, the area can heal slowly or improperly. This is why we do pain management and physical therapy. Practitioners generally describe seeing the physical aura around an injury as angry bright red or intense heat for the pain, surrounding an area of darkness or cold at the point of the actual injury.

A similar thing happens in the case of illness, although the color and light differences are more subtle and widely distributed. Illness causes a much more immediate depletion of the body's energy. If there is a fever involved, there will be a general red haze surrounding the entire field but centered in the area most seriously affected (such as the sinuses or chest in the case of a cold or flu). Where there is an infection, the color is usually described as nasty bile green or mustard yellow and the consistency as sticky or slimy.

EMOTIONS

Emotions affect the aura brilliantly and with great variation, and can change more quickly than any other energy field. In the course of normal life, the emotional aura is fairly quiet, just as most people are not intensely emotional all the time. It varies in size according to the intensity of feelings and is usually perceived around the head and shoulders. It is not always limited to a single color just as our feelings are not limited to single simple emotions, but can be seen as variously colored flower petals, wings, or flames around the upper body and heart.

The colors of the various emotional states are usually those associated with the primary emotional states of the chakras: red for anger or lust, rose for love, yellow for enthusiasm and joy, green and rose for healing and compassion, gold and purple/violet for wisdom and spiritual activity, blue for the power of the will. They darken and appear muddy when clouded with negative emotions such as sorrow, pain, and fear.

CHAKRA CENTERS AND FIELDS

The chakra centers with their surrounding fields will sometimes appear very much like a layered flower or a multi-colored egg with the body in the center. To other people they will look like a brilliantly colored chain of lights. Regardless of how they are perceived, the chakras reflect a complex integrated picture of the physical and spiritual condition of the individual. The chakras will show which energy is most predominant in that individual's approach to life and spirituality and which area of expression is weakest or blocked. A blocked or weakened center will appear smaller, duller, or dark and blotchy. Sometimes a center will be "stuck open", causing it to operate at a much higher or broader level than the others. This will drain the energy system in just the same way a ruptured water pipe will drain the entire system of water. It will appear jagged or ripped and the color will be a garish, harsh, glaring uncontrolled light, its temperature angry. Until it is repaired, none of the other centers will work properly because they are already overworking trying to make up for what is pouring out elsewhere.

It is essential to the well-being of the individual that the chakra centers be clear and balanced so that the individual can channel energy easily and freely on all levels. Consequently, it is important that healing take place on this level as well as on the physical and emotional levels, so that a true balance and health can be realized.

CREATING REALITY - THOUGHT FORMS

Remember, when a change is made in one layer, it will cause corresponding changes in all the other layers. Half of any work with energies is in the mind, both conscious and subconscious. In order to create or bring about a condition, you must image it in your mind, meditate upon it, consider it to be an accomplished fact, and create that reality in the physical world by behaving as though it were so. For example, when working with someone who is ill or has been injured seriously, it is not enough simply to deal with the energy problems brought about by the problem. An equally important part of the care you give this individual is helping them find their own sources of strength and actively encouraging them to use them. The healer must create in his or her own reality, the working image of that individual as strong and healthy, an absolute optimistic belief in the best that can be achieved.

The second step in the process is to actively discourage and weed out all those things on any level of reality that disagree and detract from the desired result. The individual receiving the healing must, in turn, help the healing process by focusing and acting upon and with these sources of strength. Sometimes this is not easy; but unless it is done, true healing will not happen.

No matter how good the individual healer may be, there is no way that he or she can be there all the time for the person they are helping. Nor should they be. As we stated before, this is not an exercise to substitute the strength and will of the healer for that of the sick or injured party. On the other hand, when you leave or if you cannot be physically present, you may wish to leave some of the healing energy behind to be with them during their recovery to give off healing energy over a duration of time. You might do this by giving them a stone or crystal of the appropriate type and color that you have charged with healing energy. On the other hand, you might want to make this even more effective by finding a charm or figurine that the individual finds particularly cheerful or comforting. This can be particularly helpful if you are working with a child who cannot really understand the healing benefits of holding the appropriate mental and emotional attitude to help the healing process along.

You can charge virtually any object with a thought, intention, or desire just as you charged your crystal. This process works on many levels at once; and while the energy patterns are operating as you have programmed them to, the image of the figure will work on the subconscious mind of the individual. This goes back to the principle of awareness levels and healing on many levels at once. Besides the obvious psychological "edge" that this practice can give the healer, there is a deeper principal at work here. When an image is used that has a particular cultural significance - such as religious figures or figures of animals, for instance - it carries with it not only the psychological impact of its content; you may also, when you charge it, call into it the power of the archetype that it represents.

An archetype is an ideal. An archetype can be defined as the ultimate core concept of something behind an idea or image. Take, for example, a common housecat sleeping in a sunny window. This individual cat has certain personal characteristics that make him (or her) unique and special and different from all other cats. However, on a spiritual, life-essence level this cat is part of a larger construct we will call Cat. Cat represents all the qualities of "catness" such as softness, grace, agility, silence, ferocity, night vision, playfulness and so on that make up our whole idea of what a cat is on all levels. It includes not only physical attributes but the whole mind set of being a cat, and extends from house cat to tiger. Cat is the archetype of which cat is a part.

By charging a stone cat with energy, you not only concentrate healing energy within the stone, but you also call into play the agility, grace, and relaxed attitude of Cat. The purpose of using such an object to facilitate physical healing is to call into action the qualities of Cat that are appropriate to the condition to be healed. A very good use for such an image might be in the case of a sports injury such as a broken leg or seriously torn tendon. This is not just a psychological ploy to act on the mind of the injured party, but an energy work as well calling upon "catness" to be part of the healing energy focused on the injured area. Using archetypes as energy forms can be a very powerful tool.

If you decide to use a stone or crystal image, it is a good idea to make sure that the subject matter it represents holds pleasant associations for the individual you are charging it for. Not everyone likes cats. Some people actively dislike them. In this case the image of a cat would not be a good one regardless of its other associations. It is very important to choose an image that your subject will react to favorably. This will go a long way to assuring that the healing energies it represents will be welcomed by them on more than one level. The mind accepts the concept of the healing; the spirit, the archetypal association; the emotions, the intent; the body, the healing energy.

KARMA

When doing healing work with others and also while doing your own meditations and energy work you will undoubtedly come upon circumstances that seem to repeat themselves over and over again. This can be true of an injury or pattern of injuries, or it may true of a life pattern or sequence of events that constantly recurs. When this repeating pattern becomes evident in a situation and especially when these circumstances refuse to be healed, or when one is healed another similar injury or situation comes right behind it, it might be well to consider how the concept of karma applies.

Karma was originally a Hindu concept referring to the idea that the spirit is reborn in body after body throughout eternity and that the circumstances of that individual's life are either rewards or punishments for deeds that person committed in a previous life or lives. This concept has gained in acceptance and is currently incorporated into a wide variety of philosophies and beliefs; but as with many Eastern concepts, there seem to be some misconceptions about it.

Karma is the means by which a spirit learns in this Universe. It is also how the spirit seeks to achieve a balance within its own nature in pursuit of this knowledge of life. It is not a system of "carrot and stick" rewards and punishments arbitrated blindly or vengefully by a panel of judges with the purpose of teaching an entity not to "screw up". It is generally thought that, in the interval of time between one lifetime and another, a spirit is shown clearly and truthfully the scope of its life and given a choice of optional circumstances so that it can learn what it needs to, can take care of ongoing or unfinished business, or can bring things more into balance with individuals or energy systems with which it has been involved. These optional circumstances may include the gender, race, national origin, physical abilities, and economic class of the body that it will inhabit, as well as many other factors that will give the entity the opportunities it needs to achieve its predetermined goals.

This accounts for the way in which an individual's life will have the same situation recurring over and over again. Many people have remarked on how their relationship patterns change only in name while the characters and situations happen over and over. In circumstances such as these there is a significant point to be understood, and the individual must find a way in which the matter can truly be put to rest. There is no "winning" or "losing" in these situations. There is only the hope of growing in wisdom, love and understanding until the situation no longer entraps us. At this point it will no longer happen or be an issue.

Some individuals choose lifetimes with tremendous barriers to overcome, such as physical or social limitations of enormous magnitude, in order to learn some lesson or to place themselves in such a position as to teach others how to triumph over similar difficulties. Or they may intend to take care of a great many problems or get finished with a great deal of unfinished business all at once. Whatever the circumstance or lesson, karma has been described as a system of energy interactions between individuals or groups. And the system of interrelationships created and sustained by those individuals.

The phrase "neutralizing karma" means to dissolve those agreements by either reaching the agreed-upon goal or by recognizing them for what they are and deciding "not to play". The real point is that these situations come to you because you have chosen them and agreed to them as part of the exercise of your own free will in causative interaction with this Universe. You may change it, if you wish, by reaching a true and clear understanding of the situation and mechanism involved.

Now this may all sound like a fancy way of saying it's all your fault and nothing you can do will get you out of it. This is not the case at all. But the truth is that you need to be aware that this is occurring and take a closer look at what is beneath the circumstances. By examining these patterns you can become aware of the mechanics of what is really happening and move in a causative present-time way to change things. You do not need endless sessions of regression therapy to reveal the true nature of what is going on. An analysis of present-time circumstances will often help tremendously. Once you have taken the situation to its elements and realized the nature of what has been occurring, you can begin a process of meditation and conscious activity to change the damaging pattern. Remember that karma is the mechanism of a process of learning and growth. You can definitely use this process for self-healing and personal expansion. If you are doing healing work in these circumstances, you may need a great deal of patience.

This seems as good a place as any to introduce some concepts related to karma and rebirth that you will hear mentioned in a variety of places and situations.

Soul Mates

Many individuals talk with genuine longing about finding their "soul-mate". There is a renewed awareness in many people of being separated and lost from another individual who in some way will make the individual feel "whole". This is motivated by deep emotional instincts generally centered in the heart chakra, and exerts a profound influence over these individuals in their daily lives. However, there needs to be some clarification of the term "soul mate" because it has more than one meaning and, therefore, more than one resolution. Many people get confused and led down unintended trails because of their misunderstanding of what they are seeking. Because of this misunderstanding they are unable to appropriately identify the energies and entities that are drawn into their lives.

One meaning of soul mate implies a polar twin - a spirit or entity that was bonded as the opposite polarity that, in many cases, was one half of a being that split away when the being entered this Universe. Many people believe that growth and further expansion are not possible without a reunion with this twin/opposite. These individuals feel incomplete and are seeking the other half in order to become whole again and continue their growth and development. These individuals will be drawn together in a variety of situations, sometimes to solve a problem on a karmic level, sometimes as siblings, sometimes as friends, sometimes but less often as parent and child, because these relationships will most often be demonstrated as the contributive partnership of equals. They are sometimes of the opposite gender from one another and sometimes of the same gender. Sometimes their association can be simply a lifetime in the company of a long-trusted and loved companion. Sometimes they are even brought together as enemies. Whatever the nature of the current relationship, it will most generally have a positive result in the strengthening of both participants as each reinforces and supports the other.

Both participants should be warned, however, to remember that the companion is also an individual and may well have different life goals and purposes, and will most certainly have brought to the new relationship a variety of knowledge and experience that has been gained during the separation of the entities. Each must be careful to look upon the other as a separate and distinct entity apart from the self; for it is easy to take the relationship for granted and not regard the other as a fully independent individual. It should also be pointed out that although an individual may sincerely feel empty or lacking or incomplete without this twin, it must be understood that each individual is complete and needs no other to complete it. Each spark of spirit is a true and complete reflection of the One that is the Prime. Polarity is a manifestation of this Universe, not a true condition of life or spirit. It is the nature of spirit to desire to be whole and complete in its expression and to seek reunion with the Prime of which it is a part and a reflection. But an individual should not believe that it is incomplete without joining to another entity. At its worst this erroneous belief will lead to unfortunate parasitic relationships that inhibit the growth and wholeness of both participants.

Another type of soul mate is an individual who seeks another being whose presence and action he or she believes is necessary to the accomplishment of an ongoing karmic lesson or task - that is, a purpose or scenario that has been repeated on a number of occasions and that the individual has intended to resolve in the present lifetime which includes specific other individuals. The individual feels incomplete without the presence of those other individuals who have previously been part of this recurring pattern. The energy of this desire and attraction will contribute to the resolution of the karmic cycle by drawing to the individual either those specific individuals or ones who are very much like them like them, and situations similar to the previous entangling ones. This process has some validity because the life lesson or purpose is one held by more than one individual and it requires the participation of the contributing individuals to resolve it. This is called a complex interaction or interactive. There is an inner longing on the part of the participants to complete and resolve the pattern or issue which can translate itself as a craving for wholeness and resolution.

So it could be said that in a single lifetime an individual may have many soul mates because these individuals may be bound together by mutual agreement and ongoing purpose. The idea of who is the individual's soul mate may also change from lifetime to lifetime, depending on the purpose for the lifetime and the relationships the individual wishes to strengthen or resolve.

Sometimes an individual may not be able to bond with the actual entities who have participated in the interactive. In this case the individual attract others whose pattern and purpose are closely similar to the individuals who have previously played the roles in the drama. The participating individual will have his or her own reasons for accepting such a position that will fulfill a karmic purpose of his or her own.

The entity most intensely sought will most often be of the opposite gender of the individual's current polarity, since this is the bonding pull that is most strongly expressed in this Universe. When these individuals find one another, they feel as though they must mate for life in order to maintain the feeling of wholeness that they discover when they encounter one another. In either case there are pitfalls and drawbacks to this form of bonding.

That entity drawn to the individual is in itself an individual with his or her own drives and purposes that may or may not be entirely in harmony with that of the individual. It is too often the case that, upon encountering such an entity, the individual in question will have needs and expectations of that entity drawn from previous lifetimes and other interactives that may not be appropriate to the current era and life course. Understandably, this can be greatly harmful to the relationship. The failure to form the relationship based upon present-time goals and considerations, the inability or unwillingness of the individual to consider the other individual as having goals and feelings of his or her own rather than being an extension/reflection of the individual's needs and purposes, can be fatal to a present time relationship. This failure will certainly not contribute to the resolution of the karmic issue.

In some instances the two individuals may even agree consciously or unconsciously to pursue the unfulfilled goals and ambitions of previous lifetimes. Although this may well be a valid life purpose - karmically speaking - there is the unseen pitfall that the individuals may end up not resolving the issue in present time, but instead merely replaying a tragedy from memory. Too great an entanglement in past issues very often leads to a dangerous lack of objectivity. In all cases, the individuals involved should bear in mind that, regardless of the parallels, the circumstances existing in present time are the ones that must be dealt with objectively and realistically and, unless these are resolved, then the situation will only continue in force. Always remember: That was then, this is now. Now is the lifetime to resolve the issue, using present-time reality and objectivity. The past may be interesting and even instructive, but it can also offer serious entrapment if present-time reality is ignored.

MACROINTERACTIVES

"Macrointeractive" is another word for a general concept called "group karma". This also refers to an incident that has occurred to a large group of individuals that is complex and has occurred, perhaps, over a long period of time that bound together the fate of that group as a whole. An example of this would be the Holocaust of World War II. In a larger sense, however, it is the mechanism that forges national or group identities or pools of consciousness. The force and duration of the incident bound the group of people together with a single shared identity and the reality of a common purpose or threat. The karma involved in this applies not only to the singular individuals involved but to the group as a whole. It achieves great force and power of its own, drawn from all the individuals that are constantly contributing to it. Once again, it is by the conscious activity and choice of the individual that this karma can be seen as a situation of learning and growth. The sense of identity with a nation, group, or cause can be a source of great strength and wisdom, but should not stand as a substitute for individual learning and growth in the present time or relied upon as an excuse not to rise above limiting circumstances.

Aɴɪᴍᴀʟs ᴀɴᴅ Rᴇɪɴᴄᴀʀɴᴀᴛɪᴏɴ

A question that is often asked is whether or not people can reincarnate as animals. The answer is, No. Although anything is possible on a spiritual level and exceptions could occur, human beings are part of a different branch of archetypal energy than animals, and exist on a different level of consciousness and awareness than animals. Generally speaking, animals do not have a distinct continuity of spiritual consciousness in the same way that human beings do but exist as parts of the group consciousness of their archetype. When an animal dies, its spirit returns to that energy pool - the archetype - as an integral part of the greater whole. This individual spiritual energy contributes its life experience to the sum of that greater whole, allowing it to grow and expand. This in turn allows the individual members of that group to adapt as a species to changing environments and conditions. It is not reborn so much as the energy is re-embodied as yet another of its kind.

This is how animals become a domesticated species. If sufficient numbers of a species are kept in an environment that does not differ excessively from their original state, their life patterns begin to change, particularly if, as a result of selective breeding, their emotional pattern becomes altered. For instance, wild cattle are tough, stringy, and highly nervous and touchy. Domestication of wild cattle took many generations of herding and breeding so that their domestic cousins finally evolved as meaty, milk-heavy, and reasonably placid. The archetype changed significantly to accommodate this differentiation of energy patterns until it is a stable and predictably different form.

A FURTHER LOOK AT HEALING AND ENERGIES

Healing is an open-ended field of exploration that can lead you into a range of skills, techniques, questions, and sometimes answers. It is a path of self-awareness and growth, and will bring up many issues as you pursue it. The techniques and concepts discussed in this chapter are only a basic few that can form the nucleus of your working body of information and skill. As you work with other people with your healing techniques, their input and comments will prove invaluable as an indication of the impact and effectiveness of the methods you are learning. Healing is an excellent subject for exploration in your journal.

CHAPTER 7
YOUR ENVIRONMENT ON MANY LEVELS

THE NATURE OF YOUR ENVIRONMENT

The quality of your environment and the way in which it affects you begins with you. Just as you have set up shielding on yourself while you are in a passive meditative state, you can adjust the nature of the shielding from passive to active so that it shields you from unnecessary and/or negative input from your surroundings as you go about your daily business. Using your intention, you can not only adjust the level and nature of what the shield filters out, but also the nature of what it projects into the surrounding environment. By using this shield as a natural extension of your intention, you will be able to broadcast whatever is needed. A good image for this is a strong light that fills a room when you enter it. This light comes from your heart and completely dispels any shadows or darkness. If an environment is "hot" from disputes and tension, your shield could broadcast "coolness". You may wish to work on cultivating a wider band of shielding that overlays your personal inner and outer one that automatically acts to improve the atmosphere around you whether you are conscious of its action or not. This is not as strange an idea as it sounds. After all, most of the negative influences in your environment come to you in subtle ways that bypass your conscious awareness. A place may look all right when you enter it, and only later will you find that you have become depressed, tense, or angry. By extending your senses you can perceive that this environment has been the scene of disputes and tensions for a long time so that those energies remain even after the participants have departed. It only makes sense to activate these subtle senses to react to these stimuli when and where they are first perceived, rather than to wait until the conscious mind realizes that something is wrong.

There are two slightly different approaches you can take with your environment, depending on how greatly you want to interact with it and how much time you want to spend there. If you are just passing through, extending your personal shielding on a conscious level to broadcast light, peace, coolness, or whatever you feel is necessary to the situation will probably be enough. But if you are planning to spend time in the room or general environment, you may want to go further. This may also be the case when you are doing healing work in a place where there has been a lengthy illness or negative emotion. An environment can become saturated with the vibrations of illness or negativity; then part of the healing process might be to clear up the general atmosphere of the place you are working in so that the minute you walk away, the individual you are working with does not immediately become "reinfected". Remember how you grounded and cleared your stones? The process is the same but on a much larger level. In the case of a room, just stand in the center and draw all the excess up through your feet and hands and then release it to where it should go.

When you are satisfied that it is cleared, replace the negative energy you have just dispersed with clear white energy. This is very important. In the last chapter, we talked about how nature hates a vacuum, and it is also true in the case of a room or environment. If you do not draw from the universal energy to fill and restore the atmosphere in a place, it will tend to fill itself with whatever is available. It may also be that the occupants of that room may subtly sense that something is "missing" and unconsciously refill the room with energy that is contrary to the healing process.

WARDED SPACE

The more sensitive you become to energies and vibrations and the more healing work you do, the more you may wish to have a private space, such as your room, workshop, or office, that has its own protective shields established. This is also a good idea if you just like peace and quiet on all levels, and if you don't want to disturb your neighbors with what you are doing. Whatever your reasons, peace and privacy have their merits; and by using creative imaging you can create a space with shields that exist whatever else you may be doing or wherever else you may be.

This operation may take some time, but remember that you are establishing a permanent base of operations. This should not have to be repeated as long as you are using the space, provided that you remember to ground both yourself and the space after you are done working. You must also remember that you will still have to do your own grounding and not rely on the space to do it for you. The space is a passive environment and will reflect what you put into it. No amount of warding or grounding will be able to keep the space peaceful and safe if you are constantly bringing negativity and strife into it.

Warding a room requires activity on two levels at once - both spiritual and physical. Because the room exists in the physical universe, you will be using physically manifested activities to energize the process, just as you will be using higher-level energies in the form of imaging and intention to dictate the quality and intensity of the energies you will be directing. Before beginning, you should take a few minutes to relax and put yourself in a calm and open frame of mind, raising your energy by cycling it between your palms. The first step is to cleanse the space in the room, removing any negative or unwanted vibrations from it. If you wish, you may use incense to cleanse the air and as an image for driving away the negativity while replacing it with healthy peaceful energy. Walk slowly, clockwise around the room, stopping at the center of each wall. Raise your hands and build a strong image of the wall of the room glowing a soft pearly white. Repeat this at each wall. Then stand in the center of the room and repeat this imaging for the ceiling and the floor, so that when you have completed the process, you are standing in a box of glowing light. Take some time at each step of the process to hold the image in your mind until it is clear and solid, until your inner mind "believes" what you are establishing. If you are not comfortable with visualizations, you can use other sensory methods to construct your room. You may want to image the walls as a temperature change - that is, warmer or cooler as a barrier to other temperature differentials coming from the outside. Or you may use the tactile image of an electrical field that zaps unwanted incoming negativity, like a bug light zaps a moth. Sound, feeling, taste, smell - any sense can be used. You may spend some time finding the system that works best for you. Know firmly that each wall, the ceiling, and the floor of your space will have exactly the same properties as your personal shield. All negativity will be deflected, while all things positive and necessary will be allowed to pass through in the form in which they are best dealt with.

This process is appropriate for any place you live or sleep. On a daily basis it will provide you with a peaceful and restorative environment in which to meditate, rest, and create. This will be a good environment, not only for your private work and meditations, but also for healing work with others. Remember that, if you do healing work in this space, you should ground and cleanse the area after each individual has left, so that no residual energies remain when you are finished to influence any new cycle of work or rest.

Manifesting Energies in the Physical World

It is important that your environment reinforces your work in the larger world, because this is one way that you can bring the other worlds into physical manifestation in the here and now. Your surrounding space contributes to you in terms of your energy and spiritual needs by shielding both you from unwanted energies from the surrounding environment and by energizing you with a concentration of forces that you may feel the need of. This can be taken a step further. Just as healing entails activity on all levels of awareness - physical, mental, emotional, and spiritual -to work effectively, other forms of energy work require a process on each of these levels in order to have a result in the physical universe.

COLOR IN YOUR ENVIRONMENT

Color is a part of your overall environment, and many studies have been done on its subtle effect on everyone every day. We have discussed color and its effects on all your levels of awareness so it is understandable how this works. Just as your eyes perceive it on a visual wavelength level, so do your subtle senses perceive color energy in other ways. For instance, a room or building that is painted completely grey will become depressing after a while. Using the color of the energy you wish to cultivate as an active part of your space and work can be a very positive step and will have many subtle benefits in your life and consciousness.

Thoughts and emotions have the power to determine the content and quality of your life. Putting this together with the ideas of color and form leads to a more complex view of how your surroundings and environment will have a constant daily effect on you, on your thoughts, and on your feelings. Very few people are able to observe all their surroundings all the time on a completely conscious level. Generally, you are aware of things as being in the background compared to what you are focused on at the moment. Regardless of the small degree of attention you pay them, these background images, are, nevertheless, being observed by your subconscious mind and on all your levels of awareness. Things that you have long ago ceased to even notice on a conscious level, are moment by moment having an impact on you. This idea goes beyond color, and has to do with other elements, or lack of them, as well, such as your perception of the physical safety and/or comfort offered by the area. It also includes such simple things as images and objects. As your inner self becomes more aware and powerful, you may feel the need to take stock and reevaluate what sort of subtle messages and energies your environment is sending.

In your home physical environment, you may well be in a position to decide the colors and objects with which you surround yourself. If you are, it is a good idea to pay attention to the physical details of your life. What colors surround you?

Are they refreshing to your eyes or are they leftovers from a time in your life that has moved into your past? Do they reinforce the quality of the energies you are concentrating on developing? What do the pictures or posters on your wall show you? They should present you with images of people and things that bring out positive feelings. This does not mean that every object in your life should constitute some sort of visual pep rally, but it does mean that you should be aware of how the elements of your life affect you. You should chose them with care.

THE FOCAL POINT - WORKING TOWARDS A GOAL

Let us take the idea of a physical space another step farther. Once you have a established your own working space, you can create a focal point of energies that you wish to concentrate on or draw more intensely into your life. It is often helpful to have physical/visual keys that are charged with the energies you are working with. This will be a collection of pictures, figures, candles, flowers, stones, or other physical objects that have the energy characteristics that you wish to bring into your life. Colors, objects, smells, sounds, and images all have the power to affect you on more than one level of awareness. By manipulating energy, color, and form you can take a definite step forward into empowering yourself and drawing upon more concrete manifestations of what you need in this life. It is as though you were building a generating station from which you could draw needed energies at will. By using these elements in a concentrated space, you are making a focal point for the energies to concentrate on this plane. These energies can be used to pursue many different goals, and this focal point can be tuned to bring you the particular energies you need to achieve your purpose.

This is your stepping-off point to move into that larger world and bring its forces to bear on improving your daily life. This can be very powerful when used for your personal healing, growth, and evolution. It can also be powerful when used to improve your material surroundings and to draw the energies to you that you feel will enrich and secure your life on a daily basis. It sets in motion the impetus and momentum for positive change and fulfillment.

Why the Focal Point Works

When you worked with your crystal, you were using a physical means to tap the higher energies. This is because of a basic law of the universe that says that nothing in any universe can behave outside the laws of that particular universe. So in order for anything or any energy to operate in this universe it has to have a physical way to do it. This is not to say that higher-plane energies don't really exist here. They very much do, but, according to the laws of this plane, they must have a physical way to manifest. As you bring together and use physical objects in your work, they will be acting on your subtle senses, reaching through to your inner awareness as well as the rational mind. This is an exercise in expression and feeling, of tuning a point of focus to energies you wish to have in your life.

Color and form can be used to create a physical location in which particular desired energies are concentrated and focused to create a kind of window through which you can more easily reach and communicate with the energies of other planes. This is very like making an interface or transfer point through which these energies can enter this universe, make themselves felt, and manifest in a more concrete way. A small niche or corner will do if you do not have a whole room or office to yourself. A group of pictures and objects chosen with care and placed in a picture frame or wall box makes a starting point from which you can focus your intention so that the impulses are reinforced and strengthened. Making a small space or focal point can serve to focus and concentrate the energy you are working with. It provides you with a place to turn to from time to time, with your meditation or just in general daily life. As you do this, make a notation in your notebook on how this progresses and how it makes you feel.

THE FOUR ELEMENTS AND YOUR LEVELS OF AWARENESS

Ideally, your focal point should have representatives of each of the four elements. By assembling the representatives of the elements, you are building the energy both of that point of focus and finding that energy point within yourself. You are reaching through your levels of awareness to the point at which they connect with the larger universe, contacting their archetype. Even though you may not have thought before that this was truly part of you, this is your first step towards tracing the point where that essence of universal energy is part of your life and consciousness. In doing this work, here is a critical point to remember. As you are a product of the life force of this universe all of its potential energies exist within you. By using this focal point you enhance and amplify your awareness of it and your relationship to it, to make it into active element of your life.

Part of the purpose of this life, is working to fulfill your own potential, to alleviate unhappiness, and to achieve your dreams and ambitions. To heal and to empower yourself is the point of this work. By seeking to expand your potential and resources and make the most of that expansion, you enrich your own life, and the lives of those around you as well.

CHOOSING YOUR GOAL

THE POWER OF FEELINGS - THE EMOTIONAL LEVEL

We discussed earlier how each level of awareness is specifically connected with, and has its own particular means for dealing with, a different plane or level of existence. When you are seeking to make a fundamental change, you need to change on all the levels or the work will not hold. Each level relates to your life and goals at a different point of vibration, and each level is addressed in a way different from the others and appropriate to itself alone. The rational mind is the vehicle for dealing with the daily physical universe. It solves its problems and addresses its issues in the conscious world. The subcon-

132

scious is the doorway that allows us to find access to the planes of dreams and higher levels of awareness and existence, where we connect with our higher selves and where our spiritual bodies have their expression. The heart, the emotional world, stands half way between the spiritual self and the rational.

Emotions serve as the mediator that communicates between the levels of awareness as well as between our conscious and subconscious minds. When reaching for a complex or fundamental change in your life, it is not enough just to change your mind, you must also have a change of heart. If this were that simple, you'd have done that already and all this would be unnecessary. A physical-universe decision changes only things in the physical world. A mental decision is very likely to remain just that. The saying "going to the heart of the matter" describes exactly what must take place. When we want to bring the higher energies to us and into physical manifestation, the rational mind must be set aside, and the subtle selves must be addressed. This is done by first addressing the emotions involved.

The first step towards a solution is recognizing what you really need by identifying the emotions involved that are basically holding the condition in place. This can be defined not in terms of the eventual physical-universe results that you want to see, but in terms of the qualities and forces you need to find within yourself in order to achieve these desired results. It is this first step where the process often goes wrong. For example, a lack of money in one's life is a result, not a cause. It is the helplessness, the anger, the feeling of entrapment and limitations, that are the real issue. Loneliness and the inability to find a suitable mate can be defined either as the physical need for sexual release or as the need to find emotional fulfillment in life. As you can see, these would require two entirely different solutions. Beauty can be defined as the harmonious combination of physical features or as an air of charisma, confidence, and health. You must take some time and care to determine the basic terms of what you are feeling before you can move ahead to achieve what you want.

It is important to identify accurately the quality and content of these feelings so that your work can proceed, because it is the feelings and their roots deep within you that have the power, not your rational conscious mind at all. At this point, keep in mind that the purpose of this is not to dwell on what you cannot be, do, or have; neither is this the occasion for a lengthy soul-wrenching analysis to delineate all of your personal shortcomings, failures, and limitations. We have been taught for so long to view life as a series of problems and obstacles that the initial analysis of the situation is likely to be a lengthy meditation on a long list of losses, lacks, and pains. This will never get you anywhere. Such meditation only focuses on the problem, enhancing it and generating the negative energy that will then feed it and reinforce your own sense of helplessness and inability to cope.

REALITY AND BELIEF - THE MENTAL LEVEL

While you are analyzing what it really is that you need, it is important to keep in mind that whatever your goal may be, it should be realistic. Realistic means something that you can BELIEVE. This is very important; because if your heart can't believe it, even though you really, really want it, it will not matter how carefully you do any of the rest of the work. There will be a secret tiny voice that just says, "No", that denies the possibility of anything coming to pass, and this denial will shoot it down before you ever see any results. You must work on a gradient within your sphere of belief. A painting is done by an accumulation of brush strokes, one at a time, until, over the process of time, the image emerges. A book is written one word at a time and section by section until it coalesces and form a continuous whole based around a central theme or concept. On a smaller scale, a sandwich is eaten by first taking one bite out of the corner. It is virtually impossible to step across the wall of disbelief to see radical change accomplished overnight. It took time for things to get the way they are, and it will take time for any meaningful or lasting change to take place. But you can begin with small changes. There is an old saying that goes, "A journey of a thousand miles is begun with but a single step." Just as though you were working with a child, you must give yourself steps that you can handle; otherwise you will get discouraged and abandon the project.

So you must begin your work on a scale that is possible for your inner self to comprehend. For instance, suppose you were living in a ghetto slum, with filth in the halls and broken plaster and no running water. And suppose your primary goal were to upgrade your surroundings - a better, safer place to live. The leap from this to a splendid palace overnight is almost impossible to conceive of - there is no real belief in that gigantic a transition. But there could be belief in a place that is clean with plumbing that works. This should be the initial focus of the work. Later on the palace can take care of itself with gradual stages.

SACRIFICE - MAKING A SPACE

The need here is to find the essence or quality needed to counteract the situation. Whatever you are feeling is a form of energy - fear, desperation, anger, confusion, pain, joy, passion, love, curiosity, interest - and energy is raw material within your grasp and ready to be changed and focused. But it requires a process of transmutation. It is true that there is no "free lunch". In order to receive, you must make an exchange. In order to transmute energy, you must have something you are willing to have changed. The Universe is whole and complete; as its microcosm and reflection, you are also whole and complete. Just as in the Universal Whole nothing can be created or destroyed, it is exactly the case with your Personal Whole. Energy can only be changed.

There is the old saying that you can't get something for nothing, and that is certainly the case here. In order to receive what you are seeking, you must offer up a fair exchange - a sacrifice. Sacrifice and offering are ancient worldwide concepts that are much misunderstood in our current culture. Their true meaning is that of exchange and reciprocity, the exchange of energies between polarities of power. In order to make a space for the desired benefit to enter, an offering or sacrifice of the appropriate size must be made. The individual must give up something that has occupied a great deal of energy and attention in order to receive the greater good. This does not imply a radical, willful destruction of existing resources to achieve some uncertain outcome, but rather, a fair exchange of energy by the removal of whatever it is that is taking up the space where you want the expected good to enter. This fair exchange could mean burdensome or unneeded materials, habits, and/or attitudes that are blocking the desired benefits from coming into your life. It could mean that the heart must be given in order to receive love in return, or that an investment of money, time, or resources is required to make a project come to life and flourish. In a spiritual sense, it could mean that in order to gain wisdom or enlightenment, we must set aside our hidden agendas and prejudices in order for the real vision and clarity to be revealed. Old ideas and concepts must be set aside in order for more effective methods to work.

Whatever the specific nature of the situation, you must set aside what is occupying that space in your life in order to fill it with a new energy or concept. This isn't often easy. After all, if it were that easy, you would have done it a long time ago. And very often, these negative undesirable things that are holding this space are masquerading as benefits, so that it is even harder to release them. For example - Your life has been too much under the control of others, and in order to deal with this your goal is to become an independent and self-directed person. You are going to have to give up the "benefits" that dependency and helplessness can bring. Such deceptive benefits could be the freedom from taking responsibility for one's own decisions and taking the consequences of being wrong or the fear of failing and being unable to take care of one's self. It is very easy to let others take charge and then blame them when things go wrong. Letting go of those benefits and that dependency can be a scary step to take.

Fear of being wrong, of being rejected, of being alone - all these are powerful motivators to keep us in the same patterns we have been following. Whatever your goal, it is absolutely important to sort through your emotional, spiritual and physical "house", to identify and release what is taking up the space of the quality or power you wish to attract. Regardless of your project or goal, the first, most important and, perhaps, the most difficult sacrifice you make must be your disbelief. Until disbelief is released, real magic of change cannot enter, and all the rest is just useless effort. You cannot have the new without releasing the old.

The size of the sacrifice determines the degree of change, and most people are unable or unwilling to make such a really huge step, to sacrifice such a truly large portion of their lives on any level. The shock of it would be too great and would undermine too many systems already in place. You must realize that these negative elements form the basis for many complex patterns of thoughts, relationships, and behavior patterns, and the subconscious mind will tell you that these are critical to your happiness and well-being, even though your conscious mind may be very well aware that this pattern is undesirable or damaging. So you must realize that the sacrifice will take place on a gradient - a gradual process - just as will your desired result unfold gradually to replace what you are releasing. And rest assured, this result will take place.

NEGATIVITY

We need to pause here and look at some other viewpoints you may encounter when beginning this type of work. When you start directing energies into externally focused work, you will undoubtedly encounter people who will tell you all the reasons why you should not do so. There are some who would tell you that it is unethical to work for your own personal gain at all. There is the school of thought that says that the Universe is constantly seeking equilibrium, and that working for personal gain in some way upsets this balance. There is another that maintains that if you work for your own personal benefit, you cause the Universe to generate a negative influence on someone else to maintain the Universal balance. There are also those who believe that there is a price to be paid for everything and that the good you draw to yourself today must eventually be paid for or balanced out by a corresponding ill at some future time. There is a school of thought that says that everything that comes into your life is your karma to deal with, and if you work to change or improve your situation, you are somehow trying to avoid owning up to what you've got coming. This is all thoroughly depressing, and it almost sounds believable.

The truth is that - with a certain few exceptions - a lot of this sort of thinking mainly comes from two sorts of people. There are those who feel less than powerful in their own lives. These rationalizations all sound like wonderful excuses not to try to improve their lives, and, if you were to succeed, it would somehow threaten their comfortable shell. It might show them that there is no reason for their self-imposed limitations and this might be an uncomfortable prospect for them to deal with. The other sort of people who will tell you this are those who want in some way to see you less powerful so that either you do not threaten them or so they can maintain their control over you. This sounds like a very cynical point of view, but we strongly suggest that before you accept anyone's belief in your duty and obligation to remain powerless, you should look closely into their personal motives. Anyone who asks you to be less than you can be or endure more than you reasonably have need to should be looked at and evaluated closely, to say the least.

The Universe is already fully formed, just as it was from the moment of its creation, with all its energies and forces in perfect balance and harmony. If this were not so, chaos would erupt, and it would immediately destroy itself. The changes, creations, and destructions we observe on a daily basis are merely transmutations of the existing forces, manifestations of their infinite varied potential to combine, separate and recombine. It is a fact of both physics and metaphysics that energy can be neither created nor destroyed - only changed. When you work for your own empowerment and enrichment, you are not "taking from" the Universal pool of energy; you are not subtracting from the store of "good" positive energy or increasing the mass of "bad" negative energy to compensate for it. Making your body stronger and more powerful does not automatically make someone else weak or diminish the Universal pool of physical power. Neither does it presuppose that if you are becoming very strong in your body now, you are somehow using up your life's ration of health and strength so that in some dim future time you will be weak and sick to balance this out and compensate for what you are taking now.

Actually, you are availing yourself of what already exists freely. You are establishing your connection to this essential energy spectrum by the means and mechanisms you were born with and that exist within you. It is and always has been a part of you just as your physical strength is, and there is no reason you should not make the most of your potential.

YOUR JOURNAL AS A TOOL

As you may have gathered, there may be more to working through a goal than just making a wish. When choosing a purpose or goal for your focal point, your journal can be an excellent tool. Your journal is your private forum for recording your impressions, your dreams, your methods and techniques. Now it can also serve as a tool for analysis. As you select a goal, write it down. As you work on it in meditation, record whatever impressions, dreams, and feelings you get regarding it. Often insights will reveal themselves in a flash of realization with many elements and ideas coming together at once. Writing down these insights can make wonderful material for further reflection at a later time. Then, once your goals are firmly in place, the next step will assembling the elements for your point of focus.

CHAPTER 8
A POINT OF FOCUS AND THE STARS

Once you have decided on your desired goal, you will need to set up an externally focused space that brings elements of that goal into physical space. This will serve as the focal point for your meditation regarding the work. In this space you will begin to concentrate the elements of energy, that is, objects and illustrations of the energy quality you wish to draw. Each of these representative objects carries the essence of that element's archetype, and by placing them in your space you are basically creating a microcosm of the energy you wish to focus on. This will mean that you will need a place to put these things. Now this space can begin as anything - the end or front of a bookshelf, a framed picture on the wall, a knick-knack shelf or trifle box. You do not need an elaborate or large space, but it should be in a place that will not be physically disturbed by others and where you can spend some private, quiet time. Your space does not even need to be inside a building, although it strengthens your attachment to the process to have your focal point in your immediate environment. If you are more comfortable out of doors, or you have a special place of meditation in a garden or glade, this can make a lovely addition to it.

ART FORMS AND FEELINGS

As you assemble your point of focus, you are doing what people have done in countless ways over the millennia to connect with the universal energies. Each age and civilization has found its own particular means of physically expressing the connection with the Infinite. In the early ages of humankind, it was almost universally believed that art was the means of communications with this Universal spectrum of energies. An image of a person, animal, or object was created in order to draw upon its energy or power. The concepts of art existing for its own sake did not become popular until rather recently in human history. Paintings and sculpture have great power to move us deeply on many levels. By incorporating form, color, and content, works of art strike chords on an almost instinctual level, moving us through our emotional levels and deeper to the source of those emotions. This bypasses reason and the opportunity for the rational mind to deny or reject the energies and feelings that these powerful images stir within us. These images free us from the civilized obligation to analyze and censor, allowing the inner self to respond. It is at this inner level that change and empowerment can take root and grow. We can only do so much with our conscious mind. If change were a conscious rational process, you would have taken steps to make those changes already. You are working to reach into your deeper levels to the wellsprings from which your life and character come, and find the connections with the archetypal forces that exist in you.

THE CONTENTS OF YOUR FOCUSED SPACE

The selection of the articles for your focal point is a very personal process. The process should be done with thought and care because each article will be representative of tuned energy that speaks to you and from you about the nature of your goal. They can be purchased at a store, received as a gift, or found in nature. You might run across a photograph or picture that directly embodies the ideal of what you wish to feel or achieve. In constructing your own space, you will be searching out elements that are uniquely significant to you and the way you feel or wish to feel about your objective. What is particular about them is that they strike a chord within you that corresponds to how you feel about achieving your goal. We talked at some length in the previous chapter about the presence of objects that signify the four elements in your focused space. Each of the elements represents a level of your awareness and your connection to the universal planes that in turn correspond to those levels. Whatever choices you make should be not only on the basis of function but of feelings and intuition.

Set aside an area - a small table or shelf will do - to assemble your collection of various articles corresponding to your goal. Flowers and candles of the appropriate color, crystals and stones, and incense along with a photograph or figurine of a deity, animal, or angel figure embodying the ideal you are seeking to achieve can all contribute strongly to drawing the specific energy you wish. This central image will be the center of your focal point, holding the essence of its content and energy, embodying the highest ideal of the goal you have set for yourself. By choosing a figure of an ancient deity, you are effectively calling upon the energy of that archetype. This is a powerful act, because by doing so you are focusing upon an ideal that has been recognized and reinforced over many thousands of years and by many thousands of people, and drawing upon the energy available in that archetype. The image itself is the trigger mechanism that will allow your inner self to reach through and connect with this source. You should look through the pantheons of the world and choose the one with which you are the most comfortable or that strikes the strongest feeling within you. If you are unfamiliar with the names, it might be helpful to find a book on world mythology.

You may be drawn to a particular culture or artistic style, or you may be drawn to a particular image. Although there are general equivalents across the world and throughout the ages, there are subtle variations of character and intent that would make some more or less appropriate for your intent than others. It might also help to consider if you are more comfortable with a male or a female representation, or if you would prefer to have both to represent the dual polarity of a quality. You may find a particular animal, either real or mythical, whose qualities and bearing call into your heart the energies you wish to attract. There are even traditionally associated angels, if you are more comfortable with that resonance of energy. You may take some time before choosing an image or it may come immediately to mind. The important thing in this is that you find a personal resonance with the archetype/ideal and that, above and beyond the ideas and definitions that the figure represents, the image touches the feelings in you most closely associated with the ideal or goal you wish to make your focal point.

After you have made this choice, it will be time to choose the other elements of your focal point that will strengthen, broaden, and reinforce this central image. It does this in two ways - by stimulating your unconscious senses to open to these energies, and also by attuning this space to be receptive to and to amplify these energies as they are available from the larger world. The very act of assembling and arranging these objects focuses and attunes your attention and energies.

The Process of Assembly

When you have found your central figure or image, you should begin to gather elements that represent each of the four elements that are also in keeping with your central idea. As you carry this idea through the process of complete assembly, you are also bringing your own consciousness, element by element, into focus on your project. Incense or potpourri of the appropriate scent is a good representative of air, in outdoor areas where the effects of incense would likely be lost, air can be represented by a windchime or wind sock, if your focal point is on a windowsill with breezes, you might choose both. Use colored candles to represent the element of fire. The color of the candle should correspond to the nature of the work you are doing. Perhaps you will find a candle holder with a special shape or design that also contributes to the content and feeling. We have discussed at length the significance that color has on the levels of awareness; and the focal point of the candle flame combined with the rich glow of the candle itself provides a powerful point of focus during your meditation. When you light a candle in your space, you signal to yourself on all levels that this is the time to concentrate and open your inner self. The flame of the candle provides the point of meditation. The energy of the flame too provides a force in and of itself. To represent water, use a colored glass of water. Choose the color of the glass just as you did the color of your candle. If you would prefer, you could use a sea shell instead or in addition. Earth can be represented by a stone or crystal of a suitable color and type, or it can be a specially charged clear quartz cluster.

When you are doing this, please take into account common sense and safety. The incense you choose should be pleasant to your nose and to the noses of those you live with. An incense or flower that makes you sneeze explosively will not help your meditations. Shop around until you have found one that is suitable to both your purpose and your nose. Candles should not be left burning unsupervised and should be placed in an appropriate dish or container that will not let the wax run everywhere and that will not break when it burns out completely. Loud rackety windchimes are useless if they distract you. But, most importantly, remember that all these things that you are assembling are not just an attachment to the energy you are working with. They are an expression of you, of how you feel about this energy. Finding and assembling these things is your first step toward an examination of these energies and their feelings as they reside within you as an extension and reflection of the larger world.

A DAILY VISIT

Once your focal point is assembled you should set aside some time each day to meditate with it. Fill a glass of the appropriate color with water and place it next to the image you have chosen. Then light the candle and the incense. As you handle each item, take a moment to focus briefly on the element it represents. Open your mind to each special energy until you can feel them all around you. Then sit quietly, focus on the central object or image, and allow your mind to receive whatever impressions and guidance come to you at this time. Sometimes it will be nothing more spectacular than a feeling of understanding or well-being, a lightening of your burden or a clarity of spirit. Sometimes you may receive images - mental pictures or impressions. You may even feel as though you are receiving information that will be helpful to you in broadening your awareness of the project you have in mind. Whatever you feel, you should write down your impressions as clearly and completely as you can along with the details of the work you are doing, what your goal is, what you have assembled, when and how you meditate and focus on your goal. After a while, a pattern will emerge of impressions and information that should prove very helpful to you in your life and in your inner journey to growth.

PLANETARY ENERGIES

Sometimes the elements of your goal may remain elusive. You are saying, in effect, "If I knew what I wanted, there wouldn't be a problem." What you need is a way to see these needs and qualities objectively - to place them in categories along with the colors and other abstract energies that most appropriately apply to them. Then you can begin to assemble an external focal point that includes them. You have observed how a single clear ray of energy, perceived as light on the physical plane, differentiates into the seven basic colors. These colors embody the key signatures of the principal energies of this plane and are reflected in each element of creation. They appear in stone types as colors and crystal families, and in our bodies as chakras. On a much larger scale, they correspond to planets. It is the planetary energies that can give you some powerful guideposts to assembling your point of focus and examining the essential quality of energy you want to connect with and manifest into your life.

The planetary energies have served humanity for eons as a way to express and describe particular qualities and powers of energies, existing in a heavenly or "otherworldly" form. Many peoples and cultures throughout history and across the world have named these planets for the gods and goddesses of their religion, ascribing the powerful higher energies that rule human destiny to these remote heavenly lights. In ancient times, the forces that rule the elements and the forces and qualities that surround and motivate us were personified as gods and goddesses (just as in English the names of the planets are also the names of the deities in the Greco-Roman pantheon). We know them as archetypes because they directly embody the basic elements of human life and behavior. They represent an ideal of a particular quality of energy.

There are seven major planets, that can be observed with the naked eye and were studied extensively in ancient times - Sun, Moon, Mercury, Venus, Mars, Jupiter, Saturn. They are put in this order because this is how they appear from the Earth, as though the Earth were the center of the Solar System. Even though science has known otherwise for a long time, astrologically and arcanely they are still counted in the old order. In the old system, the Sun and Moon are also counted as planets.

The seven ancient planets symbolize and/or embody specific categories of energy and also correspond to many other symbols, concepts, forces, and even deities. For thousands of years the planets have served as a means to personify and categorize the conditions of their existence, their needs and the forces operating in their lives as a way to better focus on them, understand and, perhaps, to deal with them. It is interesting to note that many ancient cultures, although they may have been widely separated geographically, have many similar concepts about what the planets represent.

For use as an example we will continue with the idea of becoming more independent and self-directed. You may meet some opposition to this idea, not only from others, but from yourself. You might wish to attune yourself to Mars energy for a while to reinforce your personal energy and to help remove any hidden barriers you may have within yourself that might prevent you from achieving your goal in a healthy and positive way. Mars energy is not always the military force associated with Roman Ares. There are many other images that incorporate the concept of the Mars energy while bringing other essences to bear along with it. As your central focus, you might choose an image of Athena. She was not only a self-sufficient warrior figure, but also widely praised for her wisdom, rationality, beauty, and skill at domestic arts. If you do not wish or cannot find a figure of Athena but wish to focus on her energy, you might chose the figure of an owl that was her special creature. Owls are known for their silent skill as hunters in the physical world. Symbolically they are regarded as the silent and efficient hunters of truth.

	Roman	Greek	Norse	Egyptian	Yoruba/ Santeria	Hindu
Moon	Diana Selena	Artemis		Tehuti		
Mercury	Mercury	Hermes	Odin	Thoth Anubis	Eshu Ellegua	Ganesha
Venus	Venus	Aphrodite	Freya	Hathor	Oshun	Sarasvati Lakshmi
Sun	Apollo	Apollo		Horus		Surya
Mars	Mars Athena	Minerva Ares	Tyr	Sekhmet	Ogun Oya	Kali Shiva / Agni
Jupiter	Jupiter	Zeus	Thor	Amon-ra	Chango	Ganesha
Saturn	Saturn	Chronus				Yama Durga
Earth	Hera Gaia	Demeter	Frigga	Isis		Tara
Underworld	Pluto	Hades		Osiris		
Ocean	Neptune	Poseidon	Nerthus		Olokun Yemaya	

	Moon	Mercury	Venus	Sun	Mars	Jupiter	Saturn
Color	White	Yellow	Green / Rose	Orange	Red	Purple	Blue/Black
Metal	Silver	Bronze	Copper	Gold	Iron	Tin	Lead
Stone	Moonstone Opal Beryl Pearl	Neutral Agate	Emerald Jade Malachite	Amber Topaz Heliodore	Ruby Garnet Bloodstone Jasper	Amethyst Lapis Lazuli Sapphire	Onyx Obsidion
Chakra	Crown	Solar Plexus	Heart / Basal	Belly	Base	Brow	Throat
Angel	Gabriel	Ophiel (Haniel)	Hagiel Michael	Raphael	Khamael	Tzadkiel	Tzaphkiel
Scent	Lavender Clary Sage Lime	Balsam Fir Lily of the Valley	Roses Sandalwood Ylang Ylang	Laurel Cinnamon	Dragon's Blood High John The Conquerer	Myrrh Franken-sense	Sage
Day of the Week	Monday	Wednesday	Friday	Sunday	Tuesday	Thursday	Saturday

148

MOON

INNER SELF
HEALING &
INTUITION
METAL: SILVER
COLOR: WHITE
ELEMENT:
WATER

Just as everyone has an outward facing "daytime" side that is visible to everyone around us, everyone also has an inner, secret side that is not seen and is only visible through its reflections such as moods and behavior. As the Sun corresponds to your outward face, the Moon (that is never seen by its own light, but only by reflected light) rules this inner person - feelings, intuitions, visions, and dreams. It rules the home and family, that are the expressions of the inner private person rather than one's public face. The Moon also rules the things governed by or in rhythm with its forces such as the tides and waters and the female reproductive cycles and organs.

The Moon is about the inner person. Whether you are seeking inner healing or working to expand and broaden your intuition and perception of the larger world, Moon energy is part of that pattern. The Moon offers nurturing as well as guidance and is said to rule both mothers and teachers. Go to the Moon energy when you feel the need of inner healing, when you feel entrapped by your life or unable to solve a tangled problem. Moon energy rules intuition and inner guidance and can open up your inner vision to new avenues of solutions and opportunities; because of this connection it is ascribed to the Crown chakra.

There are many practitioners who say that, because the Moon rules this inner world and the connection to higher wisdom, it should be considered when doing any working of power. It is important to consider what phase the Moon is in. The growing or "waxing" phase is said to be most powerful for drawing things to the individual, for making things come about, for work concerning healing, fertility, and increase in general. When it is growing thinner, "waning" is the time for removing things and severing ties. There is an old gardener's saying that you should weed your garden and pull up stumps when the Moon is waning so that they won't come back. This is also the case in a wider scope of activity. A good example of how this works would be if you were working for abundance, more money, more financial plenty, and security. When the Moon is waxing or full, work for all this bounty and/or the ability, knowledge, and opportunity necessary to attain or achieve it. When the Moon is waning or dark, work to have poverty and the barriers to abundance removed, for the banishing of ignorance, lack of resources, and bad luck.

Just as the Moon has many aspects or faces, so are its attributes personified by many different deities - the goddesses Artemis/Diana, the Virgin Huntress, Hecate, dark goddess of magic, and Isis, in both her capacity as Divine Wife and Mother and her role as Mistress of Magic. The Moon in its three visible phases is also used to describe and symbolize the Triple Goddess - Maiden, Mother, and Crone - power and self-sufficiency for the virgin huntress, creativity for the mother, and wisdom for the crone.

The Moon's metal is silver. It rules all white and milky stones such as moonstones, white star sapphires, opal, and diamond, and watery ocean stones such as pearl, mother-of-pearl, and coral.

☿

MERCURY

MESSENGER

METAL: BRASS

COLOR: YELLOW

ELEMENT: AIR

CHAKRA: SOLAR
PLEXUS

Mercury is often seen pictured as the winged messenger of the gods; as such he rules short journeys and swift communications, both on the mundane and psychic levels. He deals with all writings and authors, record keeping, and therefore, schools and research. His qualities are eloquence and intelligence - all capacities of the intellect. There are also many stories associating him with pranks and thefts, and he therefore rules covert activities, thieving and spying. The energy is also associated with the Opener of Ways. It is said to be Mercury that carries prayers to the gods and also their messages to humankind. By this it is meant that only through opening our intelligence to new ways of thinking can we achieve what we desire.

In his role as messenger he is called Mercury/Hermes; as record keeper and guide, Thoth/Anubis; as trickster and thief, Coyote/Loki. It is interesting to note that as a trickster, he is often involved in pranks that teach their victim by pointing out faults in planning or purpose or fatal flaws of character, pride, or miscalculations from which the victim can learn and grow. In effect, he trips us with our own untied shoelaces in order to make a point.

Mercury corresponds to the Solar Plexus Chakra, the Dancer, and refers to our inner center of personal balance. He opens the ways for self-knowledge. Mercury reminds us that when our own inner harmony is out of tune, nothing will go very far or very well. The work that we do will be the reflection of our imbalance. The metal of Mercury is quicksilver or brass, and its stones are neutral-colored ones such as onyx or agates.

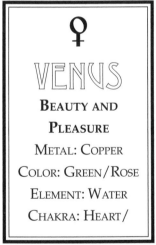

VENUS

BEAUTY AND
PLEASURE
METAL: COPPER
COLOR: GREEN/ROSE
ELEMENT: WATER
CHAKRA: HEART/

The planet Venus is associated in many cultures with the Queen of Love and Beauty; as such she embodies the aesthetic, the fertile, and the erotic. She is the source of passion. Deities that personify this energy are Aphrodite/Venus, the Egyptian cat goddess Bast, Babylonian Ishtar, Hindu Lakshmi and Sarasvati, Oshun of the Yoruba, and Norse Freya. She is the mistress of all pleasures and beautiful things as well as the people who create them - music and musicians, art and artists, and, of course, love and lovers.

The chakra center associated with Venus is the Heart - for all that is truly creative and generative derives from this source. As physical and romantic love unite male and female to the procreation of children, in a larger sense, it is the passion of Venus that unites the opposing qualities in ourselves to create and bring forth the manifestation of our highest dreams to reach

151

toward the future. It is Venus who guides the hand of the artist to bring abstract concepts into concrete reality. She represents the fertility and bounty of any project, and consequently represents money and wealth of all kinds.

The metal of Venus is copper, and her stones are green ones such as emeralds, jade, moss agate, and malachite.

OUTER PERSON

ILLUMINATION

METAL: GOLD

COLOR: ORANGE

ELEMENT: FIRE

CHAKRA: BELLY

The Sun is the ruler of outward appearances and of shedding light on what was obscured by darkness. In Greek mythology, Apollo was the driver of the fiery Sun chariot as well as the patron of the Delphic Oracle. Astrologically the Sun rules the public personality - the extroverted side of the individual. Apollo was also the patron of healers in the sense of driving out the hidden causes of disease and disorder. The Egyptian lion goddess, Sekhmet, was associated with the fiery burning principle of the Sun and was credited with the power not only to destroy the world and bring disease and plague, but also to drive out the demons of pestilence and to defend and uphold the Universal order. She was also considered the patroness of healers.

The Sun corresponds with the Warrior Chakra of the Belly. His illumination is the defense and protection against the powers of darkness and ignorance and will guard the personality - the Self. He is the bringer of truth because he drives out the shadows of deception. But be cautious about what you expose to his light. There is an old story about the boy Icarus whose father made him wings from feathers and wax. He flew magnificently through the sky, freed from prison and darkness, but he forgot to heed his father's warning. When he flew too near the Sun, the wax holding the feathers together melted and Icarus crashed to earth and was killed. The power of light and fire is double edged, and you should be aware of what you are asking when you call the Sun's energy to act in a situation.

The Sun is the sign of the Victor, and rules power and success, executives, and employers. Its influence is invoked to bring light into shadows and darkness, and to gain the favorable notice of people in high places. The Sun's metal is gold, and his stones are golden ones such as amber, and orange ones such as citrine and carnelian.

WARRIOR
METAL: IRON
COLOR: RED
ELEMENT: FIRE

Mars/Ares is the fiery red planet of the warrior energy. Other aspects of this energy are Athena/Minerva who was goddess not only of war but of learning and wisdom. Hindu Kali and Shiva in their aspects as destroyers, Agni god of fire, Ogun the Yoruba god of Iron and War, and Norse Tyr are some examples of this force. Mars energy is not simply the quality of destruction and combat. It is aggression and assertiveness; energy rather than just anger and violence. Mars clears away obstacles to progress, achievement, and self-knowledge. Mars can destroy in order to renew, and in this aspect it rules surgeons, the police, and remodeling contractors. Egyptian lion-headed Sekhmet is a warrior goddess in her aspect as Destroyer of the Enemies of World Order. Assyrian Astarte (Ishtar), goddess of war as well as Queen of Love and Beauty in her chariot pulled by lions, and Hindu Durga/Ambaji on her tiger, are excellent examples of this dual energy. Mars rules sexual energy as it is expressed in the mating instinct - the drive to reproduce and defend the young. In short, Mars rules all things that are fiery, warlike, and aggressive.

Iron is the metal of Mars, and all red stones such as jasper, garnet, and ruby.

4

JUPITER

ABUNDANCE /
NOURISHMENT

METAL: TIN

COLOR: VIOLET /
PURPLE

ELEMENT: EARTH

CHAKRA: BROW

The English words "jolly" and "jovial" meaning happy, joyful, and expansive come from the Latin spelling for Jupiter that was Jove. Jove/Jupiter/ Zeus rules abundance, growth, plenty, expansion and generosity, wealth, and prosperity of all kinds. He is in charge of speculation and money matters, increase of investments. Jupiter is described as the Father of all the gods. That means in an allegorical sense that this energy is the fount from which all the qualities and benefits of the others flow. He is associated with lightning and divine justice. The lightning of Jupiter in his role as Zeus Thunderer can be of great help to you in removing the negative effects associated with injustice such as anger, fear, grief, powerlessness, self-doubt, and illness. By helping you through this tangle of old limitations, Jupiter leads you into expansion. The energy of Jupiter is about your destiny or highest potential, and it is this influence that can show you how to be brave and forthright to go out and achieve your highest potential. Jupiter gives courage of the heart and a broader scope of vision. Like having a wise, powerful, and just father on your side, Jupiter will not do it for you, but he will show you how to do it yourself.

Jupiter's metal is tin, and his stones are blue or violet ones such as amethyst and sapphire.

ħ

SATURN

TEACHER

METAL: LEAD

COLOR: BLACK /
INDIGO

ELEMENT: AIR

CHAKRA: THROAT

Saturn is sometimes known as Father Time. As such he rules death, wills, and old things and people. As the ruler of old things and people, Saturn is the teacher of the lessons of life that include self-discipline, self-knowledge, common sense, and awareness of one's own limitations. It is the energy of Saturn that enables you to pull yourself together and get down to work. This is the energy of hard work and determination that underlies any success or achievement. The influence of Saturn is often misunderstood and is seen as gloomy, cynical, limiting, and pessimistic. But the truth is that much can be learned and much strength gained from this practical energy. Saturn is invoked in workings for protection of the hearth, home and private possessions and in this is also identified with Vesta the Roman goddess of the hearth and home. It is said that amulets of Saturn, hidden beneath the doorstep, will keep any evil from entering the home. Saturn also protects women in childbirth.

Saturn's metal is lead and his stones are black ones such as onyx, obsidian, jet, and black coral (although rulership of black coral is shared with the Moon), and indigo ones such as lapis lazuli.

ETHICS AND FEEDBACK

This is another aspect of the question of Power. Just as in healing and other energy work, the same ethical considerations apply. So before you begin the actual process of working, there are some things you need to keep in mind. The ethics of the situation dictate that you do not work against another person or for the manipulation of another individual's right to privacy and self-determination. You do not take from or act against another person. Rather, you are acting on behalf of yourself to expand and develop your own awareness and capabilities and to increase the scope of your own potentials. It is important to remember this in all the work you do. There is an old saying that goes, "You get back what you put out." We came across this principle when we talked about working with your crystal. If you put out dark angry negative thoughts and energy, that is what will come to you because that is what you are creating. The same thing is true for any other type of energy work, so it is a good idea to formulate your work in such a way that what you are creating is positive, expansive, and healthy.

AN EXPLORATION

The seven ancient planets will give you a good place to start when categorizing energies. If you are interested in doing a little personal research, it might be interesting for you to do a focal point and meditation cycle for each of them while keeping track of your impressions and intuitions in your journal. This will provide the basis for future reference when you feel that you want to do more specific work with these archetypal energies.

CHAPTER 9
CIRCLES AND GROUPS WORKING TOGETHER

POOLING YOUR RESOURCES

By bringing the four elements together in a central point of directed focus, you created a small microcosm of energy - a symbolic tiny world dedicated to the essential energy pattern that represented your goal or purpose. You constructed a tuned space dedicated and directed to one aim or goal in which you entered and worked to achieve particular benefit. But what if you wished to share that space with someone else, or with a group of people? You might wish to "pool your resources" by joining your energies together in the pursuit of some larger common goal. A group of people attuned to the same goal can be very powerful in concentrating energies.

Working with others can be a rewarding experience. It can give the work the depth and perspective that is very enriching to all the participants. Not only does group work give a wider scope and emphasis on the abilities that each participant is developing; it also provides a support group for sharing your feelings and experiences in the growth you are achieving in your private work. Group work is a good opportunity to work for broader goals than you may feel your private work is adequate to. It is also a good forum for sharing ideas, processes, and techniques that worked particularly well or images that were especially successful. Each individual has his or her unique point of view and technique on any given situation that can enrich any level of work, no matter how great or small. While working alone can build your own strength and individuality, working with others can give you encouragement and validation that will also strengthen you. Having others with whom to share your triumphs, frustrations, confusions, and insights can contribute enormously to your personal work and your expanding life in general.

THE GROUP AND ITS PURPOSE

Your group of people should be small at first - just a friend or two to start with is all that is necessary. This is so that you can become accustomed to the difference in the strength and nature of the energy that a joint effort makes. It is also a good idea to agree on a regular time to meet - perhaps once a week on a particular evening. Working with others can be a lot like learning to dance with a partner. It can be a little awkward at first until all the participants become comfortable with each other's energies and techniques.

Be sure to agree on a purpose beforehand. This purpose can vary as widely as the purpose for your solitary work. A group is a good place to explore new horizons as in guided meditation. The group can focus on a different type of energy to experiment with at each session. Each member could contribute a favorite meditation tape or write a guided session of their own to share. This will give all the members experience in energies that they might not otherwise become familiar with in the course of their private work. You can elect to set aside a time to celebrate an event or season that has particular meaning for the group's members. This can be a full moon, a season or holiday with special meaning to the group, or it can be a particularly significant event or anniversary of the group as a whole or one of its members. This may be a happy occasion such as a birthday with a guided meditation for renewal and rebirth, a successful and prosperous new year for the individual, an exploration of the possibilities and strengths another year can give. This is an energy that all the members can share and benefit from. It could also be an unhappy anniversary marking the loss of or separation from a loved one, when the group could work for healing and renewal on all levels for all the members.

It is very powerful to use group energy for healing work on a third party who may sick or injured, to send peaceful healing energy to someone who has had a recent loss, or to work for opportunity and prosperity. You can choose to work for the benefit of someone present in the group or someone who is not present, whether it is a friend, relative, or absent group member. The group can draw energy from the Universe, amplify it, and use it for the health and benefit of its members and also for the greater good of the world around you.

Group work is often a combination of several kinds of work. An example of this combination work would be choosing the time of the full moon, which is a traditional time of strength, abundance, and spiritual/psychic energy, to work on openness, prosperity and expansion for the group as a whole and its members in general, directing this strength to heal one of the members not present because of illness and sending energy to the Universe to work for world peace. When considering the group goal, you should adhere to the same guidelines as you did before. This should be something within your sphere of reality that follows ethical practice. Generating energy to help a group member find a job is an attainable objective. It is a purpose that may seem out of reach for that individual to achieve alone but is well within the sphere of strength of a group of friends working together. Seeking strength and healing for a member after a painful separation or loss is a healthy and meaningful objective for group work. Ethics should also play a part in your group work, just as it does in your private work. Finding your friend a job does not mean that someone else must lose theirs. Helping a member through a painful separation does not mean getting revenge. Remember, this is a world of plenty and opportunity for everyone.

Once the purposes and goals have been agreed upon by the group, one of the members should be chosen to guide the group, because it is helpful to have the energies coordinated at a central point of focus. This guide may be different each time you meet. If your work has more than one section, you may choose a different individual to lead each section. If someone is not comfortable leading the group, they need not do so, but it would be good for each member in turn to share the power of the focused energies of the group as a whole. The guide is responsible for giving the cues to the rest of the group so that it can work as a concerted whole and its efforts are coordinated and focused, as the conductor of an orchestra coordinates and focuses the energies of all the participating musicians. If you are working on different energies at each session it is a good idea to choose a group member to guide that session, who is particularly familiar or comfortable with the energy under study, to write a guided meditation, or to facilitate a discussion before the meditation begins. In this way, all the members of the group can profit from the experience and the unique insight of each participating individual.

EXPANDING YOUR PURPOSES

Groups are also a good format to generate and send healing energy for larger causes outside your immediate sphere of influence. As a closing exercise for your group, after the work for the immediate group is completed, it is often very rewarding to do a group meditation with a larger cause in mind - such as world peace or ethnic harmony within your community, to send love and enlightenment to world leaders. This does not have to be very specific in nature or content, but by doing this, you will be contributing to the energy of that thought form archetype. In working for a greater good, you are adding to the energy pool and archetype that contains that idea, and you can hope that eventually that energy force will become so strong that it affects the course of our physical world. You will also be joining your efforts to those of many others already doing the same sort of work, and by doing this you will also increase your share of the available working strength.

CIRCLES

Just as you created a safe space in which to do your private work, it is a good plan to create such a space for your group to work in. This safe space can be created by the group working together as a whole at each meeting. It should be your first order of official business, just as it is your own personal first order of business when working alone. Although it is the same process as creating a shining sphere of energy surrounding you, it is not quite so simple as guiding yourself through the procedure. Since all the members of the group should participate in the energy work, the procedure is somewhat more formalized, so that everyone will be aware of what is going on and what their cues are. Using these more formal procedures, the group can create a large protected area in which to move around freely, that shields and strengthens the entire group at once. These areas are called circles, although they are not flat two-dimensional circles at all. They are large, area-size globes of energy surrounding the working area of the entire group, like a huge shiny silver ball. They can be indoors or outdoors, large enough to encompass a room or large group of people, or small enough to surround just two or three people. They can be set up and taken down with each session of work, or they may be left in place on a permanent or semipermanent basis and reinforced with each session. Whatever their space, duration, or purpose, the principles behind them are the same as creating the shining sphere of energy for yourself alone. The circle space works like a combination of the warded space you established in your room and the more specific directed energies of the focal point you established.

When setting up a circle, you are creating the same kind of focal point using the same elements that you did when meditating by yourself, but on a much larger scale. You are constructing a space that is separate and apart from ordinary mundane reality. That exists, not only in the present physical universe, but in the spiritual worlds as well. You can think of it like an airport, that is, a safe departure point for your spiritual work and journeys, as well as a point of arrival for receiving energies from the other planes and levels of reality. The energy space of a circle can help you focus on the work at hand by concentrating and fine-tuning the energy you set in motion rather than allowing it to disperse and dissipate. This is similar to the way a greenhouse concentrates and focuses the light, warmth, and moisture within it to create the environment inside it.

Establishing the Circle

When working in a group it is helpful, but not always necessary, to verbalize the process and speak the elements of the process out loud. The use of verbal cues helps the members of the group coordinate the energies of their visualizations in a concerted effort. This does not mean to make a great stage production of raising the circle. It can remain as quiet, simple, and tasteful as you choose it to be.

Be sure each member of the group grounds and centers before you begin. Take a moment to become calm and peaceful - to align your own energies privately within. The group may be standing or sitting. Whatever you choose should be comfortable for all the members of the group. You may hold hands, if this is comfortable for you, in order to unify the energies. Then begin by calling upon the four elements. If there are four of you, each one can take a part, or the one you have chosen as guide can speak all the parts. You may use the physical symbols of the elements as you did with your focal point, or you may simply use a visualization of the elements. You may choose to have your guide read a longer visualized meditation or you may observe a short period of silence while each member of the group privately visualizes the essence of the element.

> "We invoke the essence of air to breathe life into our enterprise." - light the incense - Visualize the rosy pink clouds of dawn stirring with fresh breezes that refresh and awaken your awareness on all levels.

> "We invoke the essence of fire to energize and activate our aspirations." - light the candle - Visualize vibrant, living flames running all through you, bringing warmth and vitality to your body, mind, heart, and spirit.

"We invoke the essence of water to bring its intuitive guidance to our goal." - fill the bowl - Visualize sweet flowing waters brimming with life flowing all through you; see the ocean waves from which all life came, bringing life and joy to all your purposes on all levels.

"We invoke the essence of earth to give form and solidity to our goal." - hold the crystal - Visualize the abundance and solidity of the fruitful earth giving substance, strength, and duration to your work on all levels.

"We invoke the essence of the Universal Light to be here with us to manifest our goal on all levels of awareness - heart, mind, spirit, and body, united in the light of the Universe."

Visualize a clear pearly light flowing all through you and surrounding the entire group with its warmth, safety, and limitless potential.

Take another moment to ground and center with the new quality of energy. Then the guide should invoke the essence of the energy quality you have decided on and state the purpose you have come together for. You may at this point wish to share a group meditation - whether this is silent and directed to the focal point and the purpose, or whether it is a guided meditation read by one of the group members for the others to follow, is up to the group to decide. Whichever method you choose, the guide should be the one to set the pace and duration of the meditation. When the energy raised through the meditation has reached the desired level, the guide should the instruct the group to release it by sending it outward into the world to accomplish the purpose set for it. The circle does not need to be conducted entirely in a meditative state. Once you have raised it, you will find that it makes a comfortable and safe space for conversation and discussion.

CLOSURE

However simple or intricate you have made your group circle, when your work is done, you should release and dismiss the energies you have called and worked with. This will be the final release point that sends them outward to accomplish their designated goal or grounds them back into the Universal pool of energy. This can be accomplished by verbally blessing and thanking each of the elements in the reverse order in which they were called. The idea of thanking the elements serves many purposes. It recognizes that the elements have been truly present and that the contributed energy is real and valid and working on many levels toward the accomplishment of your goal. Thanking the elements also acknowledges that all of us are but a part of a large whole of consciousness in which we move and interact and for which we take responsibility in our actions and intentions on all the levels of awareness.

Once the circle is closed, each participant should take a moment to ground and center, aligning his or her energies with the energies of this physical world. This is very important. Very often group work will leave the participants feeling light-headed and euphoric. This is a lovely feeling to have for a while, but is no condition in which to drive a car or walk home. Energy must always be grounded and centered into mundane life if it is to have any lasting benefit. It is also a good idea to have some sort of food available - cookies and snacks are fine. Eating is very grounding - meat particularly so. A snack will renew those members of your group who have used more energy than they have replaced and will also help all of the participants ground and restore themselves.

Group work is a wide and rich field of exploration. It can provide a basis for personal work and a platform for a wide range of studies and experiences. As your group works together, the energy you generate will naturally draw others of a like mind who will want to participate with you. This can be a wonderful opportunity to grow. It is at this point you should be careful of how fast your group grows. When your group is first beginning, it is a good idea to keep its numbers small to control the nature and strength of the energies you are working with. Later, as you become more accustomed to working with group energy, you may wish to throw the membership wide open to take in anyone who wishes to join you for however long, or you may wish to keep the membership small and intimate. Admitting new members to a group that has been working together for any length of time can be enriching to the group as a whole, but there will always be a period of adjustment until the group members become accustomed to the new personalities and energies.

The Way On from Here

Your work, whether alone or in a group, will most definitely begin to change your world, first in small ways then in other larger ones. It can be used to achieve any outcome, whether in personal health, inner awareness, or physical well-being and abundance. You will begin to experience a system of expanded perceptions and beliefs about the nature and quality of the worlds around you that is more positive and healthy for you as a sensitive individual and spiritual being. As this system of beliefs is created and nurtured, it will begin to gather its own form of energy momentum that you may use by taking the energy of positive results achieved and using it to reinforce your expanding world.

Working with energies creates a framework of reality in which you can move and work that in its turn affects other realities (that are also frameworks of energy). You may use this to change single circumstances or events in your life, or change the pattern that your life has taken. Whatever you wish to change, it is up to you; because you have the tools and abilities within you to make those changes at any time. In the beginning of any undertaking, it will take a certain amount of intention to maintain a situation or pattern of energy, but as time goes on and that pattern becomes established, it will achieve sufficient momentum to be self-sustaining without a great deal of effort on your part. This is how your world will grow and change, created and nurtured by you. The powerful connection to the Universal energies is truly at your disposal because each one of those energies is part of you, existing within you all the time, waiting for you to develop its potential and begin to live a happier and enriching life, not only for yourself but for all the world around you.

- ABOUT THE AUTHOR -

 As both an artist and author, Maya Heath combines an abiding interest in metaphysics and higher consciousness studies with her lifelong passion for art and history. She uses her skill as an artist and art historian to fabricate images which stimulate and make an inner bridge between the seen and the unseen - the Higher and Lower aspects of the self and the other worlds which surround us. Her love of the past transforms into her vision of the future in her elegant designs for the archetypal jewelry produced by her company, Dragonscale. When not travelling, she lives with her husband, Bob, her daughter, Adrienne, spiritual extended family, many happy cats and assorted other creatures in a log cabin in the green rolling hills of the Missouri River valley in northwestern Missouri.

She is the author of The Egyptian Oracle, a unique divination system based on ancient Egyptian myths and magic. It offers a multifaceted tool for self-awareness and personal transformation. The Egyptian Oracle is published by Bear & Company.

Her most recent release from Merlyn Press, Ceridwen's Handbook of Incense, Oils and Candles, weaves together the many metaphysical elements of essential oils and aromatherapy, color, planetary and lunar influences, tarot and ritual to provide an excellent summary reference guide to the basics of ritual sympathetic magic for self-development and consciousness expansion.

Maya has appeared on television and radio and has lectured and taught on metaphysics, energy and crystal work. If you or your organization would like information regarding personal appearances, lectures, and workshops on energy work or developing your personal intuitive process; or if you would like information about archetypal jewelry and related materials, your letter can be addressed to:

ENERGIES
P. O. Box 12212
Parkville, MO 64152-0212

Or by e-mail at: energies@juno.com

The author would welcome your questions and comments on this work and will make every attempt to answer correspondence that she receives. However, due to her travel schedule and the volume of correspondence this may not always be possible, but thanks you for your time and interest.